Lecture Notes in Statistics 210

Edited by P. Bickel, P.J. Diggle, S.E. Fienberg, U. Gather,
I. Olkin, S. Zeger

For further volumes:
http://www.springer.com/series/694

Przemysław Śliwiński

Nonlinear System Identification by Haar Wavelets

 Springer

Dr. Przemysław Śliwiński
Institute of Computer Engineering, Control and Robotics
Wrocław University of Technology
Wrocław
Poland

ISSN 0930-0325
ISBN 978-3-642-29395-5   ISBN 978-3-642-29396-2 (eBook)
DOI 10.1007/978-3-642-29396-2
Springer Heidelberg New York Dordrecht London

Library of Congress Control Number: 2012949919

Printed on acid-free paper

Springer is part of Springer Science+Business Media (www.springer.com)

To my family: Tosia, Ala and Witek

Preface

„Solidarność"
The monument by E. Get-Stankiewicz

The book recounts my scientific journey through the not-yet-fully explored territory of wavelets applications in system identification, and herein I would like to express my sincere gratitude to all who influenced me with their ideas and supported during this search.

In particular I would like to sincerely thank Prof. Zygmunt Hasiewicz and Prof. Włodzimierz Greblicki for guiding me between the very first Scyllas of wavelet theory and Charybdises of nonparametric system identification.

This adventurous quest would have surely been more arduous without the help from Prof. Czesław Smutnicki, Prof. Ewaryst Rafajłowicz, Prof. Jerzy Rozenblit, and Prof. Michael W. Marcellin. I am also grateful to many traveling companions—my workmates from *Department of Control and Optimization*, Dr. Grzegorz Mzyk, Dr. Paweł Wachel, Dr. Ryszard Klempous, Dr. Ewa Szlachcic, and Dr. Jerzy Kotowski, and my colleagues from *Institute of Computer Engineering, Control and Robotics*, Prof. Ewa Skubalska-Rafajłowicz, Prof. Janusz Biernat, Prof. Stanisław Piestrak, and Dr. Krzysztof Berezowski—for countless and vigorous scientific disputes.

Finally, I would like to acknowledge the research support of my home *Wrocław University of Technology* (PWr), *Polish Ministry of Science and Higher Education* (MNiSW), and *University of Arizona*.

Wrocław, Poland
March–November 2011

Przemysław Śliwiński

Contents

Chapter 1
Introduction

Abstract In order to precisely model a real-life system or a man-made device, both nonlinear and dynamic properties of such an entity need usually to be taken into account. The generic, black-box model based on Volterra and Wiener series is capable of representing rather complicated nonlinear and dynamic interactions; however, the resulting identification algorithms are impractical, mainly due to their computational complexity. One of the alternatives, offering fast identification algorithms, is the block-oriented approach, in which systems of relatively simple (and known) structure are considered. In the book, the nonparametric identification algorithms designed for the systems from such a class are proposed, and their asymptotic and computational properties are investigated.

A majority of real-life phenomena are both nonlinear and dynamic. That is, in a given time, their outputs depend on the nonlinearly transformed present and past inputs [97, 152]. To model such relations, several algorithms, based on *the Volterra and Wiener series (black-box) approach* or on *the block-oriented one*, were proposed (see, e.g., [10, 81, 93, 123, 127, 137] and [4–7, 39, 42, 61, 74]). The former allow modeling a wide class of nonlinear systems of an a priori unknown structure, however, at the cost of a prohibitively high computational complexity of their identification algorithms (cf. e.g. [150]). In the latter, the structure of a system is known and consists of simple interconnected static (memoryless) nonlinear blocks and linear dynamic elements. One of the most exploited instances of the block-oriented systems is the Hammerstein one, which is a cascade of the input static nonlinearity followed by the dynamic linear subsystem. The importance of the Hammerstein model results from the fact that the algorithms proposed for it are computationally tractable (i.e., fast) and can easily be fine-tuned to work with several other structures, like parallel or serial-parallel, or Uryson and MISO systems (see, e.g., [6, 56, 61, 71, 72, 110]). All these systems—as natural extensions of the linear ones—can be

P. Śliwiński, *Nonlinear System Identification by Haar Wavelets*, Lecture Notes in Statistics 210, DOI 10.1007/978-3-642-29396-2_1,
© Springer-Verlag Berlin Heidelberg 2013

found in various applications (e.g., in biocybernetics [83,92,97,151,152], chemistry [79, 135, 139], control [2, 107, 140, 148], power delivery [75, 76, 82, 101], economy [13], and signal processing [40, 80, 87, 88, 102, 108, 114, 126, 155]).

Depending on the a priori knowledge available for the user, one can distinguish two main approaches to the block-oriented system identification problem:

- *Parametric*, used when models of the phenomenon are given (when, e.g., the nonlinearity is a polynomial of known degree and an order of the dynamic system is known) (see, e.g., [61])
- *Nonparametric*, applied when such a knowledge is not available, i.e., it is only assumed that the nonlinearity is bounded or integrable function and the dynamics is stable but of unknown order (see [57])

As both approaches have their strengths and weaknesses, one can easily ascertain that rather than compete with, the parametric and nonparametric algorithms complement each other: The former, if the parametric models selected a priori are correct, reduces to the problem of proper estimation of the model parameters. The resulting algorithms converge fast and thus work well even for small measurement sets. However, if the models are not correct, the algorithms suffer from the systematic (bias) error and the system characteristics remain unknown no matter how many measurements are available. In turn, the nonparametric algorithms are almost exclusively based on measurements and allow recovering the system characteristics of arbitrary shape. The price we pay for such universality is their slower convergence as a larger number of measurements is necessary to compensate the lack of initial knowledge.[1]

In the book we focus on nonparametric recovery of the nonlinear part of the system. All the presented algorithms are based on the observation made by Greblicki and Pawlak in [52] that the system nonlinearity is equal to the regression function of the system output on its input. The identification algorithms are thus the *nonparametric estimates of a regression function* adjusted to the specific conditions imposed by the system identification problem. We examine algorithms based on two types of the Haar wavelet orthogonal series: the classic, invented by Haar in [59,60], and the recently proposed unbalanced one, proposed by Girardi and Sweldens in [41]. Orthogonal series seem to be a natural choice in the nonparametric approach as they are able to represent any integrable function, while wavelets make such a representation effective (sparse). Four types of algorithms implementing the *local averaging paradigm* (see [58, Chaps. 3–8]) are examined:

- The *quotient orthogonal series* (*QOS*) algorithms
- The *order statistics* (*OS*) algorithms
- The *empirical distribution* (*ED*) algorithms
- The *empirical orthogonal series* (*EOS*) algorithms

---

[1]The third approach, *the semiparametric one*, combines the advantages of the former two (cf. e.g., [65,67]) for a general introduction and [133] for the application to system identification.

The first two are, respectively, the Haar wavelet versions of the quotient orthogonal algorithms introduced by Greblicki in [46] (and applied to Hammerstein system by Greblicki and Pawlak in [53]) and of the algorithms based on sorted measurements (order statistics) introduced by Greblicki and Pawlak in [55]. The former, in the Haar wavelet version, have also been examined by Pawlak and Hasiewicz [110] and Hasiewicz [68, 69] (see also [120]).

The latter two are the new identification algorithms. The *ED* one uses the input measurements mapped by their empirical distribution, while the *EOS* algorithm exploits the empirical orthogonal series—the unbalanced orthogonal Haar basis generated by these measurements. Both algorithms are related to the similar routines proposed in the statistical literature by Delouille et al. in [26] and by Györfi et al. in [58, Chap. 18], respectively.

The following two variants of each algorithm type are examined:

- *linear*, based on the linear approximation of the system nonlinearity and obtained from a truncated expansion series (in which the first expansion terms are used)
- *nonlinear*, in which not only the first but also the largest coefficients of the expansion are used for nonlinearity estimation [29, 30]

The nonlinear scheme is specific for the wavelet series algorithms and allows taking advantage of the compactness of the wavelet function supports (which appears to be one of their most distinguishing properties among all orthogonal series; see, e.g., [23, 96, 147]) and results in effective recovery of discontinuous system nonlinearities.

In all presented algorithms, the higher-order compactly supported wavelets can be employed instead of the Haar ones (see e.g., [27, 58, 71, 129, Chap. 18]). Nevertheless, for the following reasons, we focus on the Haar wavelets exclusively:

- *Simplicity of algorithms.* Haar wavelets are the only compactly supported orthogonal wavelets which have explicit formulas. The other wavelets are given implicitly—as the recursive procedures [24], and subsequently, they cannot be used directly in identification algorithms when the input signal is random and require designing special computational algorithms (see, e.g., [64, 66, 131, 132]).
- *Simplicity of presentation.* Identification algorithms based on Haar functions are usually of a simple form. This not only helps to demonstrate the ideas behind the algorithms and present their main properties but also greatly simplifies the corresponding proofs (cf. e.g., [16, 29, 32, 68, 69, 78, 109, 128, 147]).
- *System identification specifics.* The nonparametric identification tasks possess several peculiarities (e.g., randomness of the input signal, presence of dynamics-induced correlation of the outputs and a small signal-to-noise ratio), which, in practice, can mask the theoretical advantages of the higher-order wavelets; cf. [71, 130].

The book is organized as follows. The next chapter presents the identification problem under consideration and some examples of the Hammerstein-type systems with their unified input–output descriptions. In the third chapter, the identification goal is specified, and in the fourth, both classic and unbalanced Haar wavelets are recollected. In the fifth, the main one, the identification algorithms are proposed, and their asymptotic properties are presented. The effective computational counterparts of the algorithms are developed in Chap. 6. A summary is placed in Chap. 7. The appendix comprises all the technical materials, i.e., the proofs of theorems characterizing the convergence and convergence rates of the algorithms.

I would like to gratefully thank Prof. Zygmunt Hasiewicz from Wrocław University of Technology and Prof. Don Hong from Middle Tennessee State University for their reviews and insightful comments. I would also like to thank Dr. Eva Hiripi from Springer DE and Ms. Kumarasamy Vinodhini from SPi Global for their patience and helpful guidance throughout the manuscript preparation process.

Chapter 2
Hammerstein Systems

Abstract A discrete-time cascade of a static nonlinearity followed by a linear dynamics, i.e. the Hammerstein system, is presented. Its equivalence (from the proposed identification algorithms point of view) to some static nonlinear system with the dynamics acting as the source of an additive noise is pointed out. The ample classes of admissible memoryless nonlinear and linear dynamic elements are defined, and the assumptions concerning the input and noise signals are imposed. Selected examples of other block-oriented systems which can be described by the equivalent static system input–output equation are shown. Possible applications to high power amplifier or transmission line modeling are proposed.

We consider a discrete-time Hammerstein system (see Fig. 2.1), that is, a cascade of a nonlinear static (memoryless) block followed by a linear dynamics.

The leitmotif of the book and the main goal of the presented algorithms is to recover the nonlinear characteristics, $m(u)$, of the static part from the pairs of the system input and output measurements $\{(u_k, y_k)\}$, $k = 1, 2, \ldots$. The system is described by the input–output equation

$$y_k = \sum_{i=0}^{\infty} \lambda_i m(u_{k-i}) + z_k \tag{2.1}$$

We assume that the interconnecting signal $v_k = m(u_k)$, between the static and the dynamic part, is not available and that the system output is disturbed by an additive noise, z_k. The following proposition founds a basis for the identification algorithms considered in the book (cf. [52, 55, 57, Chap. 2]).

Proposition 2.1. *The Hammerstein system in Fig. 2.1a is equivalent to the nonlinear memoryless element in Fig. 2.1b, with the input–output equation; cf. (2.1):*

P. Śliwiński, *Nonlinear System Identification by Haar Wavelets*, Lecture Notes in Statistics 210, DOI 10.1007/978-3-642-29396-2_2,
© Springer-Verlag Berlin Heidelberg 2013

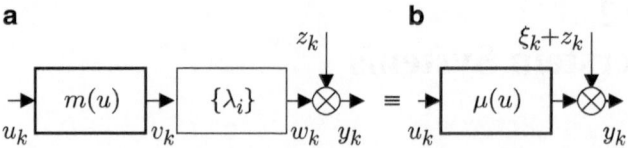

Fig. 2.1 (a) A generic Hammerstein system. (b) An equivalent static system seen from input–output viewpoint

$$y_k = \lambda_d m\,(u_{k-d}) + \sum_{\substack{i=0 \\ i \neq d}}^{\infty} \lambda_i m\,(u_{k-i}) + z_k$$

$$= \mu\,(u_{k-d}) + \xi_k + z_k, \tag{2.2}$$

in which the nonlinear system characteristics *(the* system nonlinearity*),* $\mu\,(u) = \lambda_d m\,(u) + b_d$, *where* $\lambda_d \neq 0$ *and* $b_d = E m\,(u_1) \sum_{i=0,i\neq d}^{\infty} \lambda_i$,[1] *is observed in the presence of the external noise* z_k *and the (zero-mean) system noise,* $\xi_k = \sum_{i=0,i\neq d}^{\infty} \lambda_i m\,(u_{k-i}) - b_d$.

Our a priori knowledge about the system characteristics and the signals is nonparametric, and we assume that:

1. The input, u_k, is a stationary white noise signal with a probability density function, $f\,(u)$ being Lipschitz or piecewise-Lipschitz, and strictly positive in the standardized identification interval $[0, 1]$.
2. The nonlinearity $m\,(u)$ is either a Lipschitz or a piecewise-Lipschitz function in that interval.
3. The dynamic part is linear and asymptotically stable and has the impulse response, $\{\lambda_i\}$, $i = 0, 1, \ldots$, which is of a finite or an infinite length. We assume that there is no delay in the system, i.e., $\lambda_0 \neq 0$.
4. The external noise, z_k, is any zero-mean second-order stationary signal, white or correlated.

Assumptions 1–4—being of the nonparametric nature—express rather poor prior information about the target system before the identification experiment and impose rather weak restriction on the system characteristics and on the identification conditions. In particular, the input signal can have virtually any bounded and compactly supported probability density function, viz. uniform, triangular, piecewise-constant distribution or Gauss, Cauchy, or Laplace one (truncated to the identification interval).

---

[1]Note that the multiplicative constant factor λ_d depends only on the system impulse response, while the additive second one, b_d, also on the probability density function of the input signal (i.e., on the identification conditions).

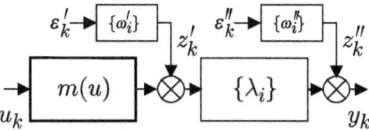

Fig. 2.2 The Hammerstein system with both the interconnecting signal v_k and the system output w_k disturbed by the external noise signals

The target nonlinearity, $m(u)$, can have isolated (jump) discontinuities. The number of jumps is unknown but finite. This assumption is satisfied by, e.g. piecewise-constant, or piecewise-Lipschitz, or, in particular, by piecewise-polynomial functions.

The next assumption, about the dynamic part, admits any discrete-time linear stable systems, that is, the systems with an absolutely summable impulse response, $\sum_{i=0}^{\infty} |\lambda_i| < \infty$. The system can thus have a finite or infinite impulse response. The response can have damped oscillations as shown in the following example:

Example 2.1. Let the dynamic system transfer function, $K_\lambda(z)$, possess only simple poles and no pole at the origin (i.e., there is no delay in the system). With ζ_r, $r = 1, \ldots, p$, denoting real and $(\eta_r, \bar{\eta}_r)$, $r = 1, \ldots, q$, denoting pairs of complex poles of $K_\lambda(z)$, the impulse response is of the well-known form

$$\lambda_n = K_\lambda(0)\delta_n + \sum_{r=1}^{p} \alpha_r \zeta_r^n + 2 \sum_{r=1}^{q} |\beta_r| \, |\eta_r|^n \cos(n\omega_r + \varphi_r)$$

where $\alpha_r = \lim_{z \to \zeta_r} (z - \zeta_r) K_\lambda(z)/\zeta_r$, $\beta_r = \lim_{z \to \eta_r} (z - \eta_r) K_\lambda(z)/\eta_r$, and $\varphi_r = \arg \beta_r$ (δ_n is the Kronecker's delta function). In particular, if there exist complex poles $(\eta_r, \bar{\eta}_r)$ or some real poles, ζ_r, are negative, then the impulse response, $\{\lambda_n\}$, includes oscillating components (in our—stable—system, all $|\zeta_r|, |\eta_r| < 1$, i.e., all poles are located within the unit circle, and the oscillations are damped).

Note that the "no-delay" assumption, $\lambda_0 \neq 0$, is made for the clarity of exposition. If $\lambda_0 = 0$, i.e., in the presence of a delay in the system, one can take any other $\lambda_d \neq 0$ and, in the following algorithms, use pairs $\{(u_k, y_{k+d})\}$ instead of $\{(u_k, y_k)\}$; cf. for comparison [57, Chap. 2.2].

As it concerns the external noise, z_k, it can—in general—be correlated and can act on both the input and the output of the Hammerstein system dynamics (cf. Figs. 2.2 and 2.5 in Example 2.5).

Example 2.2 (Multiple noise sources). Let the noise signals $\{\varepsilon_k'\}$ and $\{\varepsilon_k''\}$ in Fig. 2.2 be zero-mean, finite variance *i.i.d.* processes. The equivalent external noise is $z_k = z_k' + z_k''$ with $z_k' = \sum_{i=0}^{\infty} \sum_{j=0}^{\infty} \lambda_j \omega_i' \varepsilon_{k-(i+j)}'$ and $z_k'' = \sum_{i=0}^{\infty} \omega_i'' \varepsilon_{k-i}''$. Assumption 4 holds provided that the noise filters, $\{\omega_j'\}$ and $\{\omega_j''\}$, are stable.

Fig. 2.3 The Uryson system

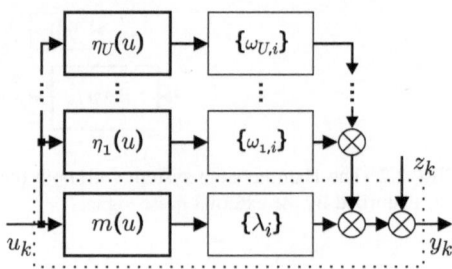

Fig. 2.3 The Uryson system

2.1 Other Systems

Several block-oriented structures can be represented in an equivalent Hammerstein system-like form, and subsequently, their nonlinearities can be recovered using the algorithms designed for the (canonical) Hammerstein systems; cf. [57, 71, 72, 110, Chap. 12]. These systems can be, in turn, used to model any phenomenon having an input nonlinearity followed by a linear dynamics of arbitrary structure, and below, several illustrative examples of such systems and circuits are demonstrated.

Example 2.3 (Uryson system). The Uryson system is an example of a multichannel nonlinear system. Its input–output equation has the following form (see Fig. 2.3 and cf., e.g., [38]):

$$y_k = \sum_{i=0}^{\infty} \lambda_i m \left(u_{k-i} \right) + \sum_{u=1}^{U} \sum_{i=0}^{\infty} \omega_{u,i} \eta_u \left(u_{k-i} \right) + z_k$$

$$= \mu \left(u \right) + \xi_k + z_k,$$

where the system nonlinearity is given by the formula

$$\mu \left(u \right) = \lambda_0 m \left(u \right) + \sum_{u=1}^{U} \omega_{u,0} \eta_u \left(u \right) + b_0$$

with $b_0 = \sum_{i=1}^{\infty} \lambda_i E m \left(u_{k-i} \right) + \sum_{u=1}^{U} \sum_{i=1}^{\infty} \omega_{u,i} E \eta_u \left(u_{k-i} \right)$, i.e., the system nonlinearity $\mu \left(u \right)$ is now a weighted sum of all nonlinearities from the system's branches (with unknown and system dependent weights). Observe however that the single nonlinearity $m \left(u \right)$ can still be separated from other nonlinearities $\eta_u \left(u \right)$, $u = 1, \ldots, U$, when the dynamics in their channels have nonzero delays (cf. Example 2.4). Moreover, when all the channel nonlinearities $\eta_u \left(u \right)$ are active (nonzero) in input signal ranges nonoverlapping with the active input range of $m \left(x \right)$, i.e., if it holds that $\text{supp} \, \mu \left(x \right) \cap \text{supp} \, \eta_u \left(x \right) = \emptyset$ for all $u = 1, \ldots, U$, then again, the $m \left(x \right)$ is separated from other nonlinearities in its activity region.

Fig. 2.4 The Hammerstein
system with a parasitic
dynamics

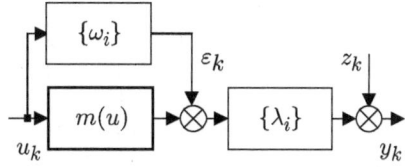

Fig. 2.5 The nonlinear
transmission line modeled
as a Hammerstein system

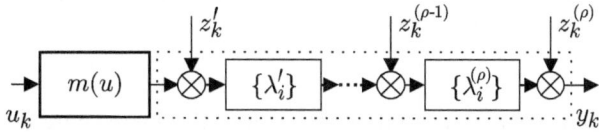

Example 2.4 (Parasitic parallel dynamics). A nonlinear system with a parallel
nuisance (parasitic, lumped) dynamics (Fig. 2.4) has the following Hammerstein
system representation:

$$y_k = \sum_{i=0}^{\infty} \lambda_i \left[m(u_{k-i}) + \sum_{i=0}^{\infty} \omega_i u_{k-i} \right] + z_k = \mu(u_k) + \xi_k + z_k,$$

where

$$\mu(u) = \lambda_0 [m(u) + \omega_0 u] + b_0,$$

with $b_0 = b_0' + b_0''$, and $b_0' = \sum_{i=1}^{\infty} [\lambda_i E m(u_{k-i}) + \lambda_0 \omega_i E u_{k-i}]$, $b_0'' = \sum_{i=1}^{\infty} \sum_{j=0}^{\infty} \lambda_i \omega_j E u_{k-i-j}$. Such system noise has a bit more complicated structure, $\xi_k = \xi_k' + \xi_k''$, where $\xi_k' = \sum_{i=1}^{\infty} [\lambda_i m(u_{k-i}) + \lambda_0 \omega_i u_{k-i}] - b_0'$ and
$\xi_k'' = \sum_{i=1}^{\infty} \lambda_i \sum_{j=0}^{\infty} \omega_j u_{k-(i+j)} - b_0''$. Note that if there is a delay in the parasitic
channel (and e.g. $\omega_0 = 0$), then we get the "memoryless system" relation

$$\mu(u) = \lambda_0 m(u) + b_0,$$

with $b_0'' = \sum_{i=1}^{\infty} \sum_{j=1}^{\infty} \lambda_i \omega_j E u_{k-i-j}$, and $\xi_k'' = \sum_{i=1}^{\infty} \lambda_i \sum_{j=1}^{\infty} \omega_j u_{k-(i+j)} - b_0''$.

Example 2.5 (Transmission line). A transmission line with an input nonlinearity
can be modeled as the Hammerstein system, see Fig. 2.5. The Assumptions 3–4
are clearly fulfilled if all noises, $z_k^{(r)}$, $r = 0, \ldots, \rho$ are zero-mean second-order
stationary processes and all the elementary components of the transmission line
$\{\lambda_i^{(r)}\}, r = 1, \ldots, \rho$, are linear and asymptotically stable dynamics; cf. Example 2.2.

Example 2.6 (Doherty amplifier). The Doherty amplifier is a nonlinear circuit used
in radio amplifiers and has recently been applied in many microwave devices (e.g., in
OFDMA-based wireless transmitters; see [20,77,88,114]). It can be seen as a system
with the input nonlinearity composed of two parallel nonlinear static subsystems

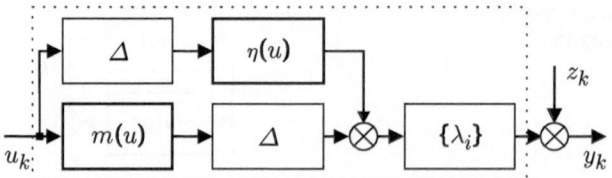

Fig. 2.6 The Doherty amplifier (Δ denotes a pure (unit) delay subsystem, i.e., $\Delta\left[m\left(u_{k}\right)\right] = m\left(u_{k-1}\right)$ and $\eta\left(\Delta\left[u_{k}\right]\right) = \eta\left(u_{k-1}\right)$)

$m\left(u\right)$ and $\eta\left(u\right)$, followed and preceded by the pure-delay elements Δ, respectively, see Fig. 2.6. Such a model has thus an equivalent, Hammerstein-like, input–output equation:

$$y_{k} = \sum_{i=0}^{\infty} \lambda_{i} \left\{\Delta\left[m\left(u_{k-i}\right)\right] + \eta\left(\Delta\left[u_{k-i}\right]\right)\right\} + z_{k}$$

$$= \sum_{i=1}^{\infty} \lambda_{i-1}\left[m\left(u_{k-i}\right) + \eta\left(u_{k-i}\right)\right] + z_{k} = \mu\left(u_{k-1}\right) + \xi_{k} + z_{k},$$

where the system nonlinearity is a (weighted) sum of both nonlinearities

$$\mu\left(u\right) = \lambda_{0}\left[m\left(u\right) + \eta\left(u\right)\right] + b_{0},$$

with $b_{0} = \sum_{i=2}^{\infty} \lambda_{i-1} E\left[m\left(u_{1}\right) + \eta\left(u_{1}\right)\right]$, and $\xi_{k} = \sum_{i=2}^{\infty} \lambda_{i-1}\left[\left[m\left(u_{k-i}\right) + \eta\left(u_{k-i}\right) - E\left[m\left(u_{1}\right) + \eta\left(u_{1}\right)\right]\right]\right]$.

2.2 Notes

The assumption about the input signal independence, while often met in the literature (see, e.g., the classical lectures by Wiener [153] and by Lee and Schetzen [93]), can clearly be pointed out as a limitation in those applications where the input signal is neither white nor can be controlled; cf. [123] and [97]. Still, in modern transmission systems, one can find the *i.i.d.* input signals being generated by the stream encoding/compressing transmitters (since a well-compressed datastream is, *in principle*, a white (and, furthermore, often of uniform distribution) process; cf. e.g. [143]).

In case when stochastic dependence of the input signal cannot be neglected, one should consider the *Wiener system* model, in which (in the simplest case) a single input nonlinearity is preceded by a linear dynamics or a *sandwich* structure, where the Wiener and the Hammerstein systems are connected in a cascade; see Fig. 2.7.

Fig. 2.7 The sandwich system being a cascade of the Wiener ($\{\omega\}$, $m(v)$) and Hammerstein systems ($m(v), \{\lambda_i\}$)

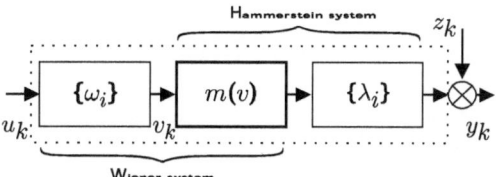

Nevertheless, it should be noted that nonparametric identification of these systems and, in particular, recovery of their nonlinearities remains a challenging problem (see the very recent results by Greblicki e.g. [48–50, 57, Chaps. 9 and 14] and [51], Pawlak et al. [111], Mzyk [105], and cf. Giri and Bai [42]).

Chapter 3
Identification Goal

Abstract The fundamental relation between the Hammerstein system nonlinearity and the regression function of the system output on the system input is presented. It allows recovery of the system nonlinearity with the help of the nonparametric regression function estimates. Several implications of this approach are discussed. In particular, the fact that the nonlinearity is estimated independently of the dynamics is emphasized. Some limitations, i.e., an ability of identification of the genuine nonlinear characteristic up to some system-dependent constants and a small signal-to-noise ratio of the measurement data, are also pointed out.

Recall that our goal is to recover the nonlinearity $m(u)$ from the input–output measurements $\{(u_k, y_k)\}$ of the whole system under the nonparametric Assumptions 1–4. The pivotal for our algorithms is the relation (2.2) in Proposition 2.1 and the resulting therefrom, well-established observation, that the regression function of the Hammerstein system output y_k on its input u_k equals to the system nonlinearity $\mu(u)$, viz., to the genuine nonlinear characteristic $m(u)$ up to some system-dependent constants λ_0 and b_0 (see [52]):

$$E\{y_k \mid u_k = u\} = \mu(u).$$ (3.1)

Proof. We have that, cf. (2.1) and (2.2):

$$E\{y_k \mid u_k = u\} = E\left\{\sum_{i=0}^{\infty} \lambda_i m(u_{k-i}) + z_k \mid u_k = u\right\}$$

$$= \sum_{i=0}^{\infty} \lambda_i E\{m(u_{k-i}) \mid u_k = u\} + E\{z_k \mid u_k = u\}$$

$$= \lambda_0 m(u) + E\{m(u_1)\} \sum_{i=1}^{\infty} \lambda_i = \mu(u),$$

since u_k and z_k are mutually independent and u_k is white and stationary. ∎

P. Śliwiński, *Nonlinear System Identification by Haar Wavelets*, Lecture Notes in Statistics 210, DOI 10.1007/978-3-642-29396-2_3,
© Springer-Verlag Berlin Heidelberg 2013

This observation associates directly the "memoryless system" input–output equation (2.2) in Proposition 2.1 with the (nonparametric) regression function estimation problem: Estimating the regression using only the system input–output measurements $\{(u_k, y_k)\}$, we are thus able to recover the Hammerstein system nonlinearity $\mu(u) = \lambda_0 m(u) + b_0$ using the nonparametric regression function estimates. Before we present our Haar wavelet-based estimates, we nevertheless will review several peculiarities of the nonparametric identification problem.

Remark 3.1. The presence of the scale and shift factors does not depend on the algorithm used to estimate the regression in (3.1). Subsequently, the nonlinear characteristic $m(u)$ cannot be, in general, recovered from the system input–output measurements under the nonparametric Assumptions 1–4 (see, e.g., [57, Chap. 2]). To find the actual $m(u)$, an additional a priori information is needed, e.g., the parametric knowledge about the subsystems, cf. [15, 104, 154], or the *active experiment* approach, viz. with a controlled input signal u_k should be applied; [14, Remark 3]. Specifically, if it is known that $m(0) = 0$ (which is often the case), then $\mu(0) = b_0$, and to recover scaled-only nonlinearity $\lambda_0 m(u)$, it suffices to estimate $\mu(u) - \mu(0)$. Similarly, if $Em(u_1) = 0$, which holds when, e.g., the input distribution is symmetric and the nonlinearity is odd, then $b_0 = 0$ and $\mu(u) = \lambda_0 m(u)$ as well (see [57, Chap. 2] and cf. [105, Chap. 8.3.3]).

The following example shows that two different systems can be tantamount from the input–output viewpoint and, in particular, can have the same system nonlinearity $\mu(u)$ (and, subsequently, the same regression function).

Example 3.1. Let the input signal u_k have a triangular distribution symmetric in the unit interval $[0, 1]$. Let the static subsystem have the characteristic $m(u) = 2 \lfloor 5(u - 1/2) + 1/2 \rfloor$, and let the dynamic part have the impulse response $\lambda_i = 1/2^{i+1}$. Let the other system have, respectively, $m(u) = \lfloor 5(u - 1/2) + 1/2 \rfloor$ and $\lambda_i = 1/2^i$. Clearly, the system nonlinearities in both cases are equal.[1]

Observe further that the nonparametric system identification Assumptions 1–4 make the regression estimation problem rather difficult. In particular, the following factors need to be taken into account during the construction (or adaptation) of the identification algorithm:

- Randomness of the input signal u_k indicates that the applicable regression estimates have to be selected from the *random setting design* class rather than from the *fixed setting design* one (where the inputs are assumed to be deterministic and (usually) equispaced); see [58, 67].
- The presence of dynamics—and its interpretation as the system noise ξ_k— implies that the random input signal is carried to the output and acts as the additional correlated system noise with an unknown distribution dependent on both the input signal probability density $f(u)$ and the shape of the nonlinearity

---

[1]The symmetric triangular *pdf* function is further used in algorithms tests (see Sect. 5.1).

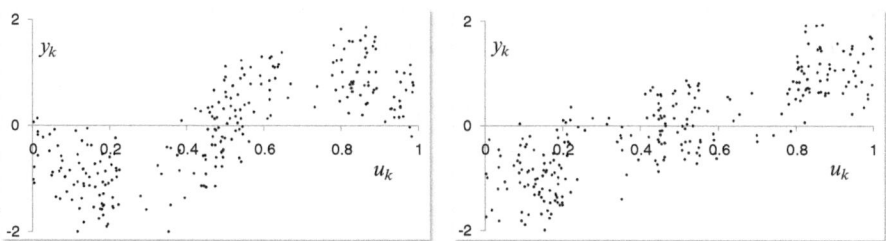

Fig. 3.1 The data points drawn from the test Hammerstein systems with either the piecewise-polynomial (*left*) or piecewise-constant (*right*) nonlinearity followed by the dynamics with the infinite impulse response $\{\lambda^i\}$, $\lambda = -1/2$ ($SNR = 1$)

$m(u)$. In result, the output measurements y_k are correlated and (even in the absence of the external noise z_k) yield a rather poor *signal-to-noise* ratio:

$$SNR = \frac{\max_u |\mu(u)|}{\max_k |\xi_k|} = \frac{\max_u |\lambda_0 m(u)|}{\max_k \left|\sum_{i=1}^{\infty} \lambda_i m(u_{k-i})\right|} = \frac{|\lambda_0|}{\sum_{i=1}^{\infty} |\lambda_i|}. \tag{3.2}$$

Clearly, the amplitude of the system noise exceeds the amplitude of the signal when $\sum_{i=1}^{\infty} |\lambda_i| \geq |\lambda_0|$. This can easily occur (for any nonlinearity) when, e.g., the system dynamics has a slowly vanishing impulse response or $|\lambda_0|$ is smaller than any $|\lambda_i|$, $i = 1, 2, \ldots$.

Finally, we would like to emphasize the fact that, under the nonparametric assumptions, the information about the system characteristics is conveyed solely by the "cloud" of noisy measurements $\{(u_k, y_k)\}$. Moreover, as it can be seen in Fig. 3.1, such unprocessed (raw) data are hardly discriminative and, prior to the proper identification routine, two different systems may appear virtually indistinguishable to a user.

3.1 Notes

The relation (3.1) between the nonlinear characteristic $m(u)$ of the static part of the Hammerstein system and the regression function $\mu(u)$ was discovered by Greblicki and Pawlak in [52] and then explored in a series of papers (see, e.g., [53–55], where various types of nonparametric regression function estimates were proposed and examined) and in their book [57].

In our book we focus on recovery of the system nonlinearity and do not consider the problem of dynamic subsystem identification. However, for the completeness of the presentation, we shortly recall the simple algorithm recovering the impulse response of the subsystem proposed in, e.g., [57, Chap. 2.3].

Proposition 3.1. *Assume that $E\{m(u_1)\} = 0$ or $E\{u_1\} = 0$ and that $E\{m(u_1)u_1\}$ $\neq 0$. For such a Hammerstein system, it holds that*

$$\text{cov}\{y_{i+1}, u_1\} = \lambda_i E\{\mu(u_{i+1})u_1\}, \quad i = 0, 1, \ldots. \tag{3.3}$$

Proof. Observe that

$$\text{cov}\{y_{i+1}, u_1\} = E\{y_{l+1}u_1\} = E\left\{\left[\sum_{l=0}^{\infty} \lambda_l m(u_{i-l}) + z_k\right] \cdot u_1\right\}$$

$$= \sum_{l=0}^{\infty} \lambda_l E\{m(u_{i-l})u_1\} = \lambda_{i+1} E\{\mu(u_{i+1})u_1\}.$$

∎

The impulse response can therefore be recovered *term-by-term* with the help of the following impulse response coefficient estimates:

$$\hat{\lambda}_i = \frac{1}{N} \sum_{k=1}^{N} y_{i+k}u_k, \quad i = 0, 1, \ldots \tag{3.4}$$

Remark 3.2. The form of the algorithm in (3.4) does not depend on the system nonlinearity. As we will see, the nonlinearity identification algorithms possess a reciprocal property, viz., they are all independent of the structure of the dynamic part. It means that both parts can be recovered independently and, in particular, that the possible inaccuracy of the one subsystem identification routine does not impact the performance of the other subsystem recovery procedure.

Chapter 4
Haar Orthogonal Bases

Abstract Two Haar wavelet bases, classic and the recently introduced unbalanced one, are presented together with the corresponding fast wavelet transforms (in the classic and lifting versions). Both linear and nonlinear (derived from the EZW algorithm) Haar approximation schemes are examined. The effectiveness of these schemes for Lipschitz and piecewise-Lipschitz functions is compared.

All the algorithms to be proposed in the remaining chapters are based on the two classes of the Haar wavelet functions:

- The classic Haar functions, discovered by Haar (see, e.g., [59, 73])
- The unbalanced Haar functions, introduced by Girardi and Sweldens in [41]

The classic Haar wavelets founded a basis for a family of orthogonal compactly supported wavelets invented by Daubechies in [22, 23] (and sometimes are referred to as *the first-generation wavelets*). The unbalanced Haar wavelets became the prototypes of *the second-generation wavelets* originated by Sweldens [142].

4.1 Classic Haar Wavelets

In this chapter (mainly based on the introduction to wavelets and multiresolution analysis (MRA) presented in [147, Chaps. 1.2.2, 3.1, 8]) we recapitulate the features of the Haar functions which are pertinent to the properties of the examined estimates. We will start with the introduction of the renowned Haar's scaling and wavelet functions and then present the resulting wavelet series representations:

- The multiscale representation
- The reproducing kernel representation
- The multiresolution representation

P. Śliwiński, *Nonlinear System Identification by Haar Wavelets*, Lecture Notes
in Statistics 210, DOI 10.1007/978-3-642-29396-2_4,
© Springer-Verlag Berlin Heidelberg 2013

Demonstrating the equivalence between them, we will further present two versions of the *fast wavelet transforms*, the classic algorithm designed by Mallat, [96, Chap. 7], and the lifting one, developed by Sweldens and having its roots in polyphase filters and multirate signal processing, cf. [24, 141] and [95, 115].

4.1.1 Multiresolution Analysis

Let V_0, a subspace of the square integrable (*finite energy*) functions space, $L_2(R)$, contain all piecewise-constant functions with (possible) jumps at integer points. All functions $\vartheta_0(x) \in V_0$ can be represented by the expansion series

$$\vartheta_0(x) = \sum_{n=-\infty}^{\infty} \alpha_{0,n}\varphi_{0,n}(x), \ n = \dots, -1, 0, 1, \dots, \tag{4.1}$$

where $\varphi_{0,n}(x) = \varphi(x-n)$, $n = \dots, -1, 0, 1, \dots$, are translations of a single function, $\varphi(x)$, called the *Haar scaling function* or the *Haar father wavelet*:

$$\varphi(x) = \chi_{[0,1)}(x), \tag{4.2}$$

where $\chi_{[0,1)}(x)$ is the index function of the right-open unit interval $[0,1)$. These translations, $\varphi_{0,n}(x)$, have compact and disjoint supports, $\operatorname{supp}\varphi_{0,n} = [n, n+1)$, and being orthonormal (i.e., satisfying the relation $\langle\varphi_{0,n}, \varphi_{0,n'}\rangle = \delta_{n,n'}$, for all $i, j = \dots, -1, 0, 1, \dots$, where $\delta_{n,n'} = \delta(n-n')$ stands for the Kronecker function), they constitute *the orthonormal basis* of the space V_0.

The expansion coefficients, $\alpha_{0,n}$, are clearly the inner products:

$$\alpha_{0,n} = \langle\vartheta_0, \varphi_{0,n}\rangle = \int_R \vartheta_0(x)\,\varphi_{0,n}(x)\,\mathrm{d}x = \int_n^{n+1} \vartheta_0(x)\,\mathrm{d}x. \tag{4.3}$$

Let now V_1 be the space of finite energy functions with jumps at half- integers. Starting again from the Haar scaling function, $\varphi(x)$, we can easily find that its scaled (twice, i.e., by the factor 2^1) and normalized (by the factor $\sqrt{2} = 2^{1/2}$) translations, $\varphi_{1,n}(x) = 2^{1/2}\varphi(2x-n)$, create an orthonormal basis of V_1. Observe also that

$$\varphi_{0,0}(x) = \varphi(x) = \varphi(2x) + \varphi(2x-1) = \sqrt{2^{-1}}\left[\varphi_{1,0}(x) + \varphi_{1,1}(x)\right]. \tag{4.4}$$

In general, the scaled and translated versions of the single Haar scaling function, defined as

$$\varphi_{mn}(x) = 2^{\frac{m}{2}}\varphi(2^m x - n), \ m = 0, 1, 2, \dots, \ n = \dots, -1, 0, 1, \dots, \tag{4.5}$$

are orthonormal

$$\int_R \varphi_{mn}(x)\,\varphi_{mn'}(x)\,\mathrm{d}x = \delta_{n,n'} \tag{4.6}$$

and constitute bases of the spaces V_m of piecewise-constant functions with jumps at binary rational points $2^{-m}n$. The supports of $\varphi_{mn}(x)$ are compact:

$$\mathrm{supp}\,\varphi_{mn} = \left[\frac{n}{2^m}, \frac{n+1}{2^m} \right). \tag{4.7}$$

Note that $V_m \subset V_{m+1}$. An infinite ladder of nested subspaces V_m, i.e.,

$$V_0 \subset V_1 \subset \cdots \subset V_{m-1} \subset V_m \subset V_{m+1} \subset \cdots$$

is the Haar *MRA* of the space $L_2(R)$. The ladder has the property that

$$L^2(R) = \overline{\lim_{m \to +\infty} V_m}. \tag{4.8}$$

4.1.2 Multiscale Approximation

An arbitrary (not necessarily piecewise-constant) function $\vartheta(x) \in L_2(R)$ can be approximated by the orthogonal projection onto the space V_K:

$$\vartheta_K(x) = \sum_{n=-\infty}^{\infty} \alpha_{Kn}\varphi_{Kn}(x) \tag{4.9}$$

with the expansion coefficients, α_{Kn}, being weighted *local averages* of $\vartheta(x)$ in the adjacent intervals $\left[2^{-K}n, 2^{-K}(n+1)\right)$, cf. (4.3):

$$\alpha_{Kn} = \int_R \vartheta(x)\,\varphi_{Kn}(x)\,\mathrm{d}x = 2^{\frac{K}{2}} \int_{\frac{n}{2^K}}^{\frac{n+1}{2^K}} \vartheta(x)\,\mathrm{d}x. \tag{4.10}$$

4.1.3 Reproducing Kernel Approximation

Inserting (4.3) into (4.1) yields the kernel representation of piecewise-constant functions, $\vartheta_0(x)$, from the space V_0:

$$\vartheta_0(x) = \int_R \vartheta_0(x) \sum_{n=-\infty}^{\infty} \varphi_{0,n}(x)\,\varphi_{0,n}(v)\,\mathrm{d}v = \int_n^{n+1} \vartheta_0(x)\,\phi(x,v)\,\mathrm{d}v$$

where

$$\phi(x, v) = \sum_{n=-\infty}^{\infty} \varphi_{0,n}(x) \varphi_{0,n}(v) = \varphi(x) \varphi(v)$$

is the *Haar reproducing kernel* in the space V_0. It has the following compact form

$$\phi(x, v) = \chi_{[0,1)}(x - \lfloor v \rfloor), \tag{4.11}$$

and its scaled versions—associated with the subspaces V_m—are defined as (cf. (4.2) and (4.11))

$$\phi_m(x, v) = 2^m \phi(2^m x, 2^m v) = 2^m \chi_{[0,1)}(2^m x - \lfloor 2^m v \rfloor). \tag{4.12}$$

For an arbitrary square integrable function $\vartheta(x)$ in a given (fixed) point x, we have the following kernel approximation formula of $\vartheta(x)$ in the approximation subspace V_K:

$$\vartheta_K(x) = \int_R \vartheta(v) \phi_K(x, v) \, dv = 2^K \int_{\frac{\lfloor x \rfloor}{2^K}}^{\frac{\lfloor x \rfloor + 1}{2^K}} \vartheta(v) \, dv. \tag{4.13}$$

Note that both approximation forms, (4.9) and (4.13), are just weighted local averages of $\vartheta(x)$. The main (and important in applications) difference between them consists in the fact that the former accommodates the global information about the whole function $\vartheta(x)$ (stored in the expansion coefficients α_{Kn}), while the latter contains the local information (around the given point x).

4.1.4 Multiresolution Representation

Assume that for some function $\vartheta_1(x)$ in the space V_1 (i.e., the space of piecewise-constant with jumps at integers and half-integers), we already have its approximation $\vartheta_0(x)$ in the space V_0 and want to get the representation in V_1.

We can compute the expansion coefficients, $\alpha_{1,n}$ from the scratch or refine the existing approximation by adding "the differences" between the approximations in V_1 and V_0; cf. [29, Sect. 3.6]. These differences belong to some detail space W_0 which are in V_1 but not in V_0. The function

$$\psi(x) = \chi_{[0,\frac{1}{2})}(x) - \chi_{[\frac{1}{2},1)}(x) = \varphi(2x) - \varphi(2x - 1) \tag{4.14}$$

is called the *Haar wavelet* (or the *Haar mother wavelet*), and its translations, $\psi_{0,n}(x) = \psi(x - n)$, $n = \ldots, -1, 0, 1, \ldots$, constitute an orthonormal basis of the detail space W_0 Fig. (4.1). These translated functions have also compact and disjoint supports, supp $\psi_{0,n} = [n, n + 1)$. It can be verified that since for all n

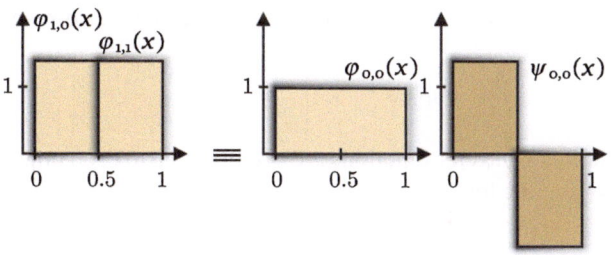

Fig. 4.1 The Haar functions family. The scaled translations $\varphi_{1,0}(x)$ and $\varphi_{1,1}(x)$ of the *father wavelet* $\varphi_{0,0}(x) = \varphi(x)$ and the *mother wavelet* $\psi_{0,0}(x) = \psi(x)$

$$\int_R \psi_{0,n}(x)\,\mathrm{d}x = 0, \tag{4.15}$$

then for all translation indices, i, j, the scaling functions, $\varphi_{0,i}(x)$, and wavelets, $\psi_{0,j}(x)$, are orthogonal, i.e.,

$$\int_R \varphi_{0,i}(x)\,\psi_{0,j}(x)\,\mathrm{d}x = 0,$$

and the detail space W_0 is orthogonal to V_0. That is, W_0 is the orthogonal complement of V_0 in V_1:

$$V_0 \oplus W_0 = V_1. \tag{4.16}$$

The representation of $\vartheta_1(x)$ has thus two equivalent forms:

$$\vartheta_1(x) = \sum_{n=-\infty}^{\infty} \alpha_{1,n}\varphi_{1,n}(x) = \sum_{n=-\infty}^{\infty} \alpha_{0,n}\varphi_{0,n}(x) + \sum_{n=-\infty}^{\infty} \beta_{0,n}\psi_{0,n}(x) \tag{4.17}$$

where $\beta_{0,n}$ are the expansion coefficients of $\vartheta_1(x)$ in the wavelet space W_1:

$$\beta_{0,n} = \langle \vartheta_1, \psi_{0,n}\rangle = \int_R \vartheta_1(x)\,\psi_{0,n}(x)\,\mathrm{d}x.$$

The relation (4.16) holds for any approximation spaces V_m and V_{m+1}:

$$V_m \oplus W_m = V_{m+1}, \tag{4.18}$$

that is, for all m, the finer approximation space V_{m+1} is an orthogonal sum of the coarser space V_m and the corresponding wavelet space W_m. The detail subspaces W_m are spanned by the scaled and translated versions of the single Haar wavelet function:

$$\psi_{mn}(x) = 2^{\frac{m}{2}}\psi(2^m x - n), \; m = 0, 1, 2, \ldots, \; n = \ldots, -1, 0, 1, \ldots. \tag{4.19}$$

The supports of $\psi_{mn}(x)$ are compact and the same as for the corresponding scaling functions $\varphi_{mn}(x)$:

$$\text{supp}\,\psi_{mn} = \left[\frac{n}{2^m}, \frac{n+1}{2^m}\right).$$

Clearly, such defined wavelet functions are orthonormal to each other and also to all scaling functions at the lower scales m, i.e.,

$$\int_0^1 \psi_{mn}(x)\,\psi_{m'n'}(x)\,dx = \delta_{m,m'} \cdot \delta_{n,n'}, \tag{4.20}$$

$$\int_0^1 \varphi_{mn}(x)\,\psi_{m'n'}(x)\,dx = 0, \text{ for } m' \geq m. \tag{4.21}$$

Combining now (4.8), (4.18), and (4.20), we obtain that the whole space $L_2(R)$ can, for any integer M, be split into the sum of the approximation space V_M and the infinite orthogonal sum of details spaces W_m, $m = M, M+1, \ldots$

$$L^2(R) = \overline{V_M \oplus W_M \oplus W_{M+1} + \cdots} = V_M \oplus \overline{\bigoplus_{m=M}^{\infty} W_m}.$$

Subsequently, the scaling functions $\varphi_{Mn}(x)$ and wavelets $\psi_{mn}(x)$, $m = M, M + 1, \ldots$ compose an orthogonal basis of the entire space $L_2(R)$ (see, e.g., [147, Chap. 1.2.2]) and can now represent (and not only approximate—as in multiscale and kernel representations (4.9) and (4.13)) an arbitrary function $\vartheta(x) \in L_2(R)$ by the following *multiresolution series* consisting on the (initial) approximation in the space V_M, some M, and on the details from the increasing scale (resolution) spaces W_m, $m = M, M+1, \ldots$; cf. (4.17):

$$\vartheta(x) = \sum_{n=-\infty}^{\infty} \alpha_{Mn}\varphi_{Mn}(x)$$

$$+ \sum_{n=-\infty}^{\infty} \beta_{Mn}\psi_{Mn}(x) + \sum_{n=-\infty}^{\infty} \beta_{M+1,n}\psi_{M+1,n}(x) + \cdots$$

$$= \sum_{n=-\infty}^{\infty} \alpha_{Mn}\varphi_{Mn}(x) + \sum_{m=M}^{\infty}\sum_{n=-\infty}^{\infty} \beta_{M+1,n}\psi_{M+1,n}(x) \tag{4.22}$$

where the scaling function and wavelet expansion coefficients, α_{Mn} and β_{mn}, are defined as

$$\alpha_{Mn} = \int_{\frac{n}{2^M}}^{\frac{n+1}{2^M}} \vartheta(x)\,\varphi_{Mn}(x)\,dx \text{ and } \beta_{mn} = \int_{\frac{n}{2^m}}^{\frac{n+1}{2^m}} \vartheta(x)\,\psi_{mn}(x)\,dx.$$

4.1.5 Equivalent Representations and Fast Wavelet Transform

For a given scale K and fixed x, all three wavelet approximations are equivalent, i.e.,

$$\vartheta_K(x) = \sum_{n=-\infty}^{\infty} \alpha_{Kn} \varphi_{Kn}(x) \tag{4.23}$$

$$= \int_R \vartheta(v) \phi_K(x, v) \, dv \tag{4.24}$$

$$= \sum_{n=-\infty}^{\infty} \alpha_{Mn} \varphi_{Mn}(x) + \sum_{m=M}^{K-1} \sum_{n=-\infty}^{\infty} \beta_{mn} \psi_{mn}(x). \tag{4.25}$$

The equivalence between the multiscale and kernel approximations, (4.23) and (4.24), has already been demonstrated. To verify the equivalence between the multiscale and multiresolution approximations, (4.23) and (4.25), we will use easy-to-verify (and well-known) relations between the scaling function and wavelet coefficients at the adjacent scales $m-1$ and m (cf. (4.4) and (4.14)):

$$\alpha_{m-1,n} = \tfrac{1}{\sqrt{2}} (\alpha_{m,2n} + \alpha_{m,2n+1}) \text{ and } \beta_{m-1,n} = \tfrac{1}{\sqrt{2}} (\alpha_{m,2n} - \alpha_{m,2n+1}). \tag{4.26}$$

Starting now from the multiscale approximation in (4.23), i.e., from the set of scaling function coefficients, α_{Kn}, at the scale K, we will get the multiresolution one in (4.25) by repeating (4.26) for $m = K, \ldots, M$, and storing the wavelet coefficients, β_{mn}, $m = K-1, \ldots, M$, together with the final scaling function coefficients α_{Mn}.

Remark 4.1. The procedures in (4.26) are the steps of the Mallat's forward *fast wavelet transform* for the Haar wavelet expansion (see, e.g., [96, Chap. 7.3]).

Conversely, to obtain the multiscale representation from the multiresolution one, we need to reverse the *fast wavelet transform* routine: Given the scaling function coefficients, α_{Mn}, and the wavelet coefficients, β_{mn}, at the scales $m = M, \ldots, K-1$, we need to compute the intermediate coefficients, α_{mn}, for $m = M, \ldots, K-1$, and store those at the final scale $m = K$. Solving to this end, the system of equations given in (4.26) with respect to $\alpha_{m,2n}$ and $\alpha_{m,2n+1}$, we get:

$$\alpha_{m,2n} = \tfrac{1}{\sqrt{2}} (\alpha_{m-1,n} + \beta_{m-1,n}) \text{ and } \alpha_{m,2n+1} = \tfrac{1}{\sqrt{2}} (\alpha_{m-1,n} - \beta_{m-1,n}). \tag{4.27}$$

Remark 4.2. The procedures in (4.27) are the steps of the Mallat's inverse *fast wavelet transform* for the Haar wavelets (see, e.g., [96, Chap. 7.3]).

4.1.6   Lifting Transform for Haar Wavelets

The fast wavelet transform procedures have an equivalent form which allows performing the wavelet transform computations in situ, that is, without an additional memory.

Assume that we have the scaling function coefficients, α_{Kn}, at the scale K. In the first step, we compute the difference (wavelet) coefficients, $\beta_{m-1,n}$, at the scale $K-1$ using the standard forward transform routine in (4.26) and then put the results in place of the odd coefficients $\alpha_{m,2n+1}$. Observing then that $\alpha_{m,2n+1} = \alpha_{m,2n} - \sqrt{2}\beta_{m-1,n}$, and inserting the right-hand side of the equation into the formula for $\alpha_{m-1,n}$ in (4.26), we obtain the following two-step sequence cf. ([24]):

$$\beta_{m-1,n} = \frac{1}{\sqrt{2}}\left(\alpha_{m,2n} - \alpha_{m,2n+1}\right) \text{ and } \alpha_{m-1,n} = \frac{1}{\sqrt{2}}\alpha_{m,2n} - \beta_{m-1,n}, \qquad (4.28)$$

which is the Haar wavelet implementation of the generic forward *lifting wavelet transform* developed by Daubechies and Sweldens [24, 141].

Remark 4.3. Computations of the scaling and wavelet coefficients, β_{mn} and α_{mn}, are, respectively, referred to as *prediction* and *update* steps of the lifting transform [141].

The corresponding *inverse lifting wavelet transform* can be obtained by simple reversing the sequence in (4.28), i.e.,

$$\alpha_{m,2n} = \sqrt{2}\left(\alpha_{m-1,n} + \beta_{m-1,n}\right) \text{ and } \alpha_{m,2n+1} = \alpha_{m,2n} - \sqrt{2}\beta_{m-1,n}. \qquad (4.29)$$

4.1.7   Convergence

Intuitively, the larger the scale factor K, the better approximation accuracy of $\vartheta(x)$ by $\vartheta_K(x)$. In this section, we formally justify this intuition examining both the pointwise and the global convergence properties of Haar approximations. The first type of convergence will be useful during exploration of the global convergence properties of the Haar estimates used in identification algorithms when the target functions are piecewise-smooth.

4.1.8   Pointwise Convergence

The fact that the function $\vartheta(x)$ is square integrable does not imply that its series expansion converge pointwise. In fact an additional smoothness condition needs to be satisfied. Namely, the Haar series converges in every point of continuity of $\vartheta(x)$ as stated by the following lemma in which the kernel approximation form (4.24) is used; cf. [147, Prop. 1.4]:

Lemma 4.1. *Let $\vartheta(x)$ be a piecewise-continuous function. Its approximation, $\vartheta_K(x)$, converges to $\vartheta(x)$ with growing K in every point of its continuity:*

$$\lim_{K \to \infty} \vartheta_K(x) = \lim_{K \to \infty} \int_R \vartheta(v) \phi_K(x, v) \, dv = \vartheta(x).$$

Proof. The proof is simple yet illustrative. Let $\vartheta(x)$ be continuous in x. For any $\varepsilon > 0$, there exists a K such that we have $|\vartheta(x) - \vartheta(v)| < \varepsilon$ when $|x - v| \leq 2^{-K}$. The Haar reproducing kernels, $\phi_K(x, v)$, are compactly supported, and

$$\vartheta_K(x) = 2^K \int_{\frac{\lfloor x \rfloor}{2^K}}^{\frac{\lfloor x \rfloor + 1}{2^K}} \vartheta(v) \phi\left(2^K x, 2^K v\right) dv = 2^K \int_{\frac{\lfloor x \rfloor}{2^K}}^{\frac{\lfloor x \rfloor + 1}{2^K}} \vartheta(v) \, dv = \vartheta(v_K)$$

for some $v_K \in \text{supp} \, \phi_K(x, v)$—by virtue of the *mean value theorem* (see e.g., [1, Chap. 3]). Since $|x - v_K| \leq 2^{-K}$, then $|\vartheta(x) - \vartheta_K(x)| < \varepsilon$, and the lemma holds.[1] ∎

By using the representations equivalence argument from Sect. 4.1.5, the lemma can immediately be applied to the multiscale and multiresolution forms (4.23) and (4.25).

4.1.9 Convergence at Jumps

In this section we analyze an interesting (and rather nonintuitive) behavior of the Haar series in discontinuity (jump) points of $\vartheta(x)$. Denote by $H(x)$ a step function, that is, a function such that $H(x) = 1$ for $x \geq 1$ and $H(x) = 0$ for $x < 0$. Observe that any piecewise-continuous function can be split into a continuous part, $\vartheta_C(x)$, and a piecewise-constant one, $\vartheta_{PC}(x)$, the latter being a sum of shifted and weighted step functions:

$$\vartheta(x) = \vartheta_C(x) + \vartheta_{PC}(x) \text{ where } \vartheta_{PC}(x) = \sum_{j=1}^{J} h_j H(x - v_j)$$

where J is a number of jumps located at points v_j and where h_j is the heights of the jth jump. Because of compactness of the Haar function supports, we can consider each jump point separately. The next lemma describes the behavior of the Haar expansion in such points; cf. [43, Th. 2.4]:

---

[1] The continuity assumption can be relaxed. In fact, Haar series converges in all Lebesgue points of $f(x)$, [85].

Lemma 4.2. *Let v be a point of a jump discontinuity of ϑ. The Haar approximation series converges to $\vartheta\,(v_+)$, that is, to the right-hand limit of ϑ, when v is a binary rational number.*

Proof. We already know that the Haar expansion converges to $\vartheta_C\,(x)$ in all points and will focus on the piecewise-constant part $\vartheta_{PC}\,(x)$. Assume for simplicity that $\vartheta\,(x)$ is just a single step function with a jump in v, i.e., let $\vartheta\,(x) = \vartheta_{PC}\,(x) = H\,(x - v)$, then

$$\vartheta_K\,(v) = \int_R \phi_K\,(v, u)\,H\,(u - v)\,du = 2^K \int_v^{\frac{\lfloor 2^K v \rfloor + 1}{2^K}} du = 1 - \left(2^K v - \lfloor 2^K v \rfloor\right).$$

For any binary rational v, there exists a scale K_0 such that $2^K v = \lfloor 2^K v \rfloor$ for all $K \geq K_0$, and hence $\lim_{K \to \infty} \vartheta_K\,(v) = 1 = \vartheta\,(v_+)$. Otherwise, the term $2^K v - \lfloor 2^K v \rfloor$ oscillates endlessly, and $\vartheta_K\,(v)$ has no limit. ∎

Example 4.1. Let $\vartheta\,(x) = H\,(x - 1/3)$, then $\vartheta_K\,(1/3) = 2/3$ for even and $\vartheta_K\,(1/3) = 1/3$ for odd Ks.

One can now easily ascertain that such a behavior (i.e., divergence of the series at nonbinary rational jump points) is caused by the discrete (dyadic) nature of the Haar expansion: The support of $\phi_K\,(x, v)$ shifts with continuously changing v in the discrete manner (viz., with a dyadic step $2^{-K} \lfloor 2^K v \rfloor$), and the convergence properties are no longer shift invariant.

4.1.10 Convergence Rate

In the previous section the conditions of the pointwise convergence of Haar expansion were established. Here, we examine the rate of this convergence, that is, the rate, the following approximation error:

$$|\vartheta\,(x) - \vartheta_K\,(x)| \tag{4.30}$$

vanishes for a given x, where $\vartheta_K\,(x)$ is of either of the equivalent approximations in (4.23)–(4.25) and where $\vartheta\,(x)$ is a Lipschitz function.

Definition 4.1. A function $\vartheta\,(x)$ is Lipschitz if it satisfies the Lipschitz condition, i.e., if there exists a constant $c > 0$, such that

$$|\vartheta\,(x) - \vartheta\,(v)| \leq c\,|x - v|. \tag{4.31}$$

Example 4.2. Any differentiable function with a bounded derivative is Lipschitz continuous. A piecewise-polynomial function with separate jump discontinuities is Lipschitz continuous between jumps.

To establish the approximation error, we will use the multiscale approximation form in (4.23). Because the compact supports of $\varphi_{Kn}(x)$ do not overlap, we have for any fixed x that

$$\vartheta(x) - \vartheta_K(x) = \vartheta(x) - \sum_{n=-\infty}^{\infty} \alpha_{Kn}\varphi_{Kn}(x) = \vartheta(x) - \alpha_{Kn}\varphi_{Kn}(x)\big|_{n=\lfloor 2^K x \rfloor}$$

and, by virtue of the *mean value theorem*,

$$\alpha_{Kn} = \int_R \vartheta(v)\,\varphi_{Kn}(v)\,dv = 2^{\frac{K}{2}} \int_{\frac{n}{2^K}}^{\frac{n+1}{2^K}} \vartheta(v)\,dv = 2^{-\frac{K}{2}}\vartheta(v_K),$$

for some v_K, which belongs to the support of the corresponding $\varphi_{Kn}(x)$. Hence,

$$\vartheta_K(x) = \alpha_{Kn}\varphi_{Kn}(x)\big|_{n=\lfloor 2^K x \rfloor} = 2^{\frac{K}{2}} \cdot 2^{-\frac{K}{2}}\vartheta(v_K) = \vartheta(v_K)$$

and

$$|\vartheta(x) - \vartheta_K(x)| \le |\vartheta(x) - \vartheta(v_K)|.$$

Eventually, since $|x - v_K| \le 2^{-K}$, we get[2]:

$$|\vartheta(x) - \vartheta_K(x)| \le c\,|x - v_K| \le c2^{-K}.$$

We have shown the following lemma:

Lemma 4.3. *If $\vartheta(x)$ is Lipschitz in the neighborhood of some point x, then the pointwise error of its Haar approximation (4.23)–(4.25) decays there exponentially with the growth of the approximation scale K:*

$$|\vartheta(x) - \vartheta_K(x)| \le c2^{-K}. \tag{4.32}$$

Intuitively, the smoother the approximated function, the faster should be the convergence rate; (see e.g., [96, Chap. 6.1]). Nevertheless, in the book, we consider only Haar approximations for which the rate in (4.32) is the best possible even for smoother, e.g., multiple differentiable, functions. That the best for Lipschitz functions rate $\mathcal{O}(2^{-K})$ does not accelerate further with a growing smoothness of $\vartheta(x)$ is a well-known property of the Haar wavelet approximation (cf., e.g., [29, Sect. 3.1]) and explains why we can confine our analysis to the functions from the Lipschitz class.

---

[2]In the book, we use the common symbol c, $c > 0$, to denote all generic constants.

Remark 4.4. If $\vartheta(x)$ is piecewise-constant, then for all continuity points, there exists a scale K_0 such that for all $K \geq K_0$, we have

$$|\vartheta(x) - \vartheta_K(x)| = 0.$$

Piecewise-constant functions are examples of *trivial functions*, for which the approximation error is smaller than the otherwise best possible one established in Lemma 4.3, and are subject of a *saturation theorem* in approximation theory (see, e.g., [29, 31, Sect. 3.1]).

4.1.11   Global Convergence

Examining the global (integrated) approximation error, we assume that $\vartheta(x)$ has a compact support and is defined in the unit interval $[0, 1]$. We will first consider the classic *linear approximation* (nonadaptive) scheme which is equivalent to the multiscale approximation. Then, we will propose and examine two *nonlinear (adaptive) approximation* schemes, in which only selected expansion terms are added to the multiscale approximation.

Remark 4.5. In the *linear scheme*, two approximants, $\vartheta'_K(x)$ and $\vartheta''_K(x)$, are in the same approximation space V_K, and so there is their linear combination, $\alpha\vartheta'_K(x) + \beta\vartheta''_K(x)$. It is not the case in general for nonlinear approximations. Moreover, the term *nonlinear approximation* expresses the fact that in these schemes the selection of expansion terms to be included in the approximant, $\vartheta_K(x)$, depends on the target function $\vartheta(x)$; cf. again [29, Sect. 2].

4.1.12   Linear Approximation

We start the analysis with uniformly Lipschitz functions $\vartheta(x)$ in the interval $[0, 1]$. Then, we will study the approximation error for piecewise-Lipschitz functions. We will exploit the multiscale approximation form of $\vartheta_K(x)$ but also will make use of its multiresolution representation.

The Haar expansion was originally designed by A. Haar as the orthogonal basis of the space $L_2[0, 1]$; see [59, 60]. Since we assume that $\vartheta(x)$ is square integrable in $[0, 1]$, this implies that

$$\lim_{K \to \infty} \int_0^1 [\vartheta_K(x) - \vartheta(x)]^2 \, dx = 0, \tag{4.33}$$

where (cf. (4.23) and (4.25))

$$\vartheta_K(x) = \sum_{n=0}^{2^K-1} \alpha_{Kn}\varphi_{Kn}(x) \tag{4.34}$$

$$= \sum_{n=0}^{2^K-1} \alpha_{Mn}\varphi_{Mn}(x) + \sum_{m=M}^{K-1}\sum_{n=0}^{2^m-1} \beta_{mn}\psi_{mn}(x). \tag{4.35}$$

Moreover, since the following version of Parseval's identity holds

$$\int_0^1 \vartheta^2(x)\,dx = \sum_{n=0}^{2^K-1} \alpha_{Kn}^2 + \sum_{m=K}^{\infty}\sum_{n=0}^{2^m-1} \beta_{mn}^2, \tag{4.36}$$

then the global (squared) error of the approximation of $\vartheta(x)$ by $\vartheta_K(x)$ can be expressed in terms of wavelet coefficients β_{mn}, from all the scales $m = K$, $K+1,\ldots$, which not incorporated in the approximation $\vartheta_K(x)$; cf. (4.35) and (4.36):

$$\text{ISE }\vartheta_K = \int_0^1 [\vartheta_K(x) - \vartheta(x)]^2\,dx = \int_0^1 \left[\sum_{m=K}^{\infty}\sum_{n=0}^{2^m-1} \beta_{mn}\psi_{mn}(x)\right]^2 dx$$

$$= \sum_{m=K}^{\infty}\sum_{n=0}^{2^m-1} \beta_{mn}^2. \tag{4.37}$$

To find a bound for this error, we need to compute a bound for the wavelet coefficients. Using the compactness of the support of $\psi_{mn}(x)$ and the vanishing moment (4.15) arguments, we get

$$\beta_{mn} = \int_R \vartheta(v)\,\psi_{mn}(v)\,dv = \int_{\frac{n}{2^m}}^{\frac{n+1}{2^m}} \vartheta(v)\,\psi_{mn}(v)\,dv$$

$$= \int_{\frac{n}{2^m}}^{\frac{n+1}{2^m}} [\vartheta(v) - \vartheta(u)]\,\psi_{mn}(v)\,dv,$$

which, for Lipschitz functions, yields

$$|\beta_{mn}| \le c2^{\frac{m}{2}} \int_{\frac{n}{2^m}}^{\frac{n+1}{2^m}} |\vartheta(v) - \vartheta(u)|\,dv$$

$$\le c2^{-\frac{m}{2}} |v - u| \le c2^{-\frac{m}{2}}2^{-m}. \tag{4.38}$$

Inserting (4.38) into (4.37), we obtain the lemma which characterizes the rate of the global convergence of the Haar approximation:

Lemma 4.4. *Let $\vartheta(x)$ be square integrable in $[0, 1]$. The integrated error of the approximation of $\vartheta(x)$ by $\vartheta_K(x)$ vanishes exponentially with a growth of the approximation scale K (cf. 4.32):*

$$\text{ISE}\,\vartheta_K \leq c \sum_{m=K}^{\infty} \sum_{n=0}^{2^m-1} \left(2^{-\frac{m}{2}} 2^{-m}\right)^2 = c \sum_{m=K}^{\infty} 2^{-2m} \leq c2^{-2K}.$$

Let now $\vartheta(x)$ be a piecewise-Lipschitz function with an unknown arbitrary but finite number of (separate) jump discontinuities. Denote the number of jumps by q. Using the compactness support argument, we observe that, at each scale m, each jump intersects with the support of only one wavelet function $\psi_{mn}(x)$ and hence impacts only one wavelet coefficient β_{mn}. In particular, if the jump is at a point x, then the affected coefficients at the scale m are the one with a translation factor $n = \lfloor 2^m x \rfloor$.

The wavelet coefficients satisfy there the following inequality, cf. (4.38):

$$|\beta_{mn}| = \int_{\frac{n}{2^m}}^{\frac{n+1}{2^m}} |\vartheta(v)\,\psi_{mn}(v)|\,dv = 2^{\frac{m}{2}} \int_{\frac{n}{2^m}}^{\frac{n+1}{2^m}} |\vartheta(v)|\,dv$$

$$\leq c2^{-\frac{m}{2}} \tag{4.39}$$

Moreover, since the jumps are separated, then there exists a scale K such that for all scales $m = K, K+1, \ldots$, each jump belongs to the support of a different wavelet function $\psi_{mn}(x)$. Splitting the approximation error, $\text{ISE}\,\vartheta_K$, into two parts corresponding to the smooth and jump regions:

$$\text{ISE}\,\vartheta_K = \sum_{m=K}^{\infty} \sum_{n=0}^{2^m-1-q} \beta_{mn}^2 + \sum_{m=K}^{\infty} \sum_{n=1}^{q} \beta_{mn}^2,$$

and inserting there the respective bounds (4.38) and (4.39)

$$\text{ISE}\,\vartheta_K \leq c \sum_{m=K}^{\infty} \sum_{n=0}^{2^m-1-q} 2^{-3m} + c \sum_{m=K}^{\infty} \sum_{n=1}^{q} 2^{-m},$$

we get the following lemma:

Lemma 4.5. *Let $\vartheta(x)$ be a piecewise-Lipschitz function in the interval $[0, 1]$. The integrated approximation error vanishes exponentially with the growing scale K and*

$$\text{ISE}\,\vartheta_K \leq c2^{-2K} + c2^{-K} \leq c2^{-K}. \tag{4.40}$$

Observe that the rate in (4.40) is the best possible rate for linear (nonadaptive) approximation schemes; cf. [18, Th. 2].

Remark 4.6. In case of the integrated approximation error, the class of trivial functions, i.e., those for sufficiently large-scale K, the approximation error is zero

$$\text{ISE}\,\vartheta_K = 0,$$

consists of all piecewise-constant functions with jumps in binary rational points, i.e., of all functions from any approximation space V_K; cf. Remark 4.4 and see [29, Sect. 3.1].

From Lemma 4.5 and Remark 4.6, one can immediately conclude the following interesting corollary:

Corollary 4.1. *Let $\vartheta\,(x)$ be a piecewise-Lipschitz function in the interval $[0, 1]$ with an arbitrarily number of jumps located in dyadic (binary rational) points. Then, the asymptotic convergence rate of the Haar approximation is the same (optimal) as for uniformly Lipschitz functions, that is,*

$$\text{ISE}\,\vartheta_K = \mathcal{O}\left(2^{-2K}\right).$$

The nonlinear approximation scheme presented in the next chapter allows to achieve the same optimal rate also for piecewise-Lipschitz functions with jumps located in arbitrary points.

4.1.13   Nonlinear Approximation

In order to improve the approximation accuracy in a piecewise-Lipschitz case, we propose the nonlinear approximation scheme derived from the *heuristic EZW (embedded zero-tree wavelet)* algorithm invented for image compression; see [125] and cf. [122].

EZW Approximation Scheme

In this chapter we assume that q, the number of discontinuities in $\vartheta\,(x)$, is known.[3] In principle, the EZW scheme consists in adding to the linear approximation (cf. (4.34) and (4.35)) the sets Q_m the q wavelet coefficients β_{mn}, located in the

---

[3]This assumption is further removed, and in all identification algorithms, it will be assumed that the actual number of discontinuities is unknown (but finite).

Fig. 4.2 In the EZW
nonlinear approximation
scheme, the wavelet
coefficients β_{mn},
$m = M, \ldots, K-1$ added to
the linear model with the
scale M are located in the
cones of influence generated
by the discontinuity points x_1
and x_2 of $f(x)$

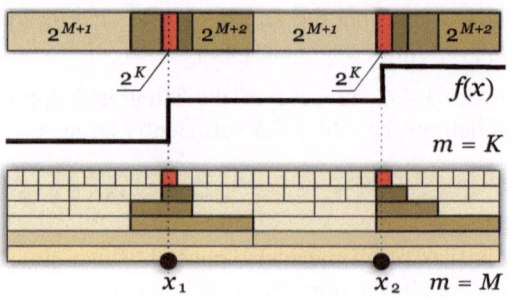

cones of influence of the jump points (rather than adding *en masse* the entire set of
the wavelet coefficients residing at the scale $m = M, \ldots, K-1$). The concatenation
of these sets,

$$Q_{MK} = \bigcup_{m=M}^{K-1} Q_m,$$

consists thus of the wavelet coefficients which approximation error is of a order
$\mathcal{O}\left(2^{-m/2}\right)$ at each additional scale $m = M, \ldots, K-1$; cf. (4.39). These
coefficients—presented in the wavelet domain—form the *discrete cones of influence*
induced by the jump points of $\vartheta(x)$ (cf. e.g. [96, Chap. 9.2] and Fig. 4.2).

In order to find a bound for the resulting approximation error, consider again the
error bound in (4.40) and observe that for piecewise-Lipschitz functions, the latter
term, $c2^{-K}$, corresponding the approximations error of the jump part, is of a larger
order than the former one, $c2^{-2M}$, and is responsible for a slower decay of the order
of the overall approximation error, ISE ϑ_K. Using this observation, one can propose
the following approximant (cf. (4.25)):

$$\vartheta_{MK}(x) = \sum_{n=0}^{2^M-1} \alpha_{Mn}\varphi_{Mn}(x) + \sum_{m=M}^{K-1}\sum_{n \in Q_m} \beta_{mn}\psi_{mn}(x) = \vartheta_L(x) + \vartheta_{NL}(x)$$

$$(4.41)$$

consisting of the *linear (nonadaptive)* multiscale approximation $\vartheta_L(x)$ and of
the *nonlinear (adaptive)* part $\vartheta_{NL}(x)$, with the wavelet coefficients β_{mn} terms
corresponding to the wavelet function $\psi_{mn}(x)$ whose supports include jumps. The
approximation error is now

$$\text{ISE}\,\vartheta_{MK} \leq \sum_{m=M}^{\infty}\sum_{n=0}^{2^m-1} \beta_{mn}^2 + \sum_{m=K}^{\infty}\sum_{n \in Q_m} \beta_{mn}^2 \leq c2^{-2M} + c2^{-K}.$$

Selecting the scale K such that both linear and nonlinear error components, $c2^{-2M}$
and $c2^{-K}$, are of equal orders, viz., taking $K = 2M$, we obtain the lemma; cf.
Lemma 4.5 and [96, Prop. 9.4].

Lemma 4.6. *If $\vartheta(x)$ is piecewise-Lipschitz, then the approximant in (4.41) for the approximation scales, M and K, selected such that*

$$K = 2M,$$

approximates $\vartheta(x)$ with the error

$$\text{ISE}\,\vartheta_{MK} \le c2^{-2M}. \tag{4.42}$$

Comparing the errors (4.40) and (4.42), one can observe that:

- The order of the error bound, $c2^{-2M}$, being the best possible for a Lipschitz functions, cannot be improved and adding to ϑ_{NL} terms at the scales beyond the scale $K = 2M$ does not decrease the order of the overall error, ISE ϑ_{MK}.
- The nonlinear approximant in (4.41) consists of

$$2^M + q(K - M) = 2^M + qM$$

terms, while the linear approximant (4.23) with the same order of approximation error would have $2^K = 2^{2M}$ terms.
- Conversely, given a budget of $2^M + qM$ terms (which, for large scales M, remains of order $\mathcal{O}(2^M)$), the linear approximant yields the error of order $\mathcal{O}(2^{-M})$, while the nonlinear one offers the smaller error order $\mathcal{O}(2^{-2M})$.

Example 4.3. For the same order of accuracy, the linear approximant requires

$$\frac{2^{2M}}{2^M + qM} = \mathcal{O}(2^M)$$

times more terms than the nonlinear one, i.e., for $\vartheta(x)$ being piecewise-Lipschitz, application of the nonlinear approximant results in $\mathcal{O}(2^M)$ less terms.

N-Term Approximation

The applied *EZW* technique is a version of the well-known N-term nonlinear approximation scheme (see, e.g., [16–18, 29, 96]) and can be found in, e.g., [96, Chap. 9.2]. In this approximation scheme, the terms $\beta_{mn}\psi_{mn}(x)$ with the *actually* largest coefficients β_{mn} are included in the approximant, while in the *EZW* algorithm, the terms whose coefficients are *potentially* the largest (i.e., whose coefficients bounds have the largest orders) are selected.

Remark 4.7. Both approaches are asymptotically equivalent since, for sufficiently large-scale M, all the wavelet coefficients β_{mn} located inside the cones of influence have larger values than those outside (because of their slower vanishing rate, $\mathcal{O}(2^{-m/2})$ vs. $\mathcal{O}(2^{-m})$).

4.2 Unbalanced Haar Wavelets

The classic Haar functions are generated by the scaled translations of a pair of father
and mother wavelets $\varphi(x)$ and $\psi(x)$, and both the scale and translation factors,
m and n, are known a priori (deterministic). The resulting functions $\varphi_{mn}(x)$ and
$\psi_{mn}(x)$, $n = 0, \ldots, 2^m - 1$, are thus translated copies of themselves. This makes
the Haar functions well adapted to *fixed design* problems where the input data
are deterministic and equidistant (e.g., in time series and image processing) but
rather poorly performing in *random design* problems where the inputs are randomly
distributed. To address this problem, the design-adapted Haar functions (referred
to as *unbalanced Haar wavelets* or *empirical Haar orthogonal series*) have been
proposed in [41]. The functions are not known a priori but defined accordingly to
the given inputs $\{u_k\}$, $k = 1, \ldots, N$.

We shortly recollect their construction (which appears to be related to order
statistics formalism). Assume for simplicity that all inputs are from the unit interval
and that N is a *dyadic integer*. Denote further by $\{x_k\}$ a sorted (ordered) version
of the inputs $\{u_k\}$, $k = 1, \ldots, N$. The elements x_k create a random partition
of the unit interval consisting of disjoint and adjacent intervals (*sample blocks*),
$\chi_k = [x_{k-1}, x_k)$, $k = 1, \ldots, N$. Let $I_k = |\chi_k| = x_k - x_{k-1}$ denote the spacing,
i.e., the lengths of the intervals χ_k; cf. [25]. The scaling functions constituting an
orthonormal base of the finest approximation space are directly generated by these
measurements

$$\varphi_{\log_2 N, k}(x) = \sqrt{I_k^{-1}} \chi_k(x),$$

where $\chi_k(x)$ denotes the indicator function of the sample block interval $\chi_k = [x_{k-1}, x_k)$.

In general, for any $m = 0, \ldots, \log_2 N$, the normalized *unbalanced scaling
functions* are defined as follows:

$$\varphi_{mn}(x) = \sqrt{I_{mn}^{-1}} \chi_{mn}(x), \tag{4.43}$$

where $\chi_{mn}(x)$ are indicator functions of the intervals χ_{mn} defined as concatenations
of the adjacent sample blocks

$$\chi_{mn} = \bigcup_{k=(n-1)2^L}^{n2^L - 1} \chi_k,$$

with $L = \log_2 N - m$ and $n = 1, \ldots, 2^m$ and where I_{mn} is the length of χ_{mn}

$$I_{mn} = |\chi_{mn}| = \sum_{k=(n-1)2^L}^{n2^L - 1} I_k, \tag{4.44}$$

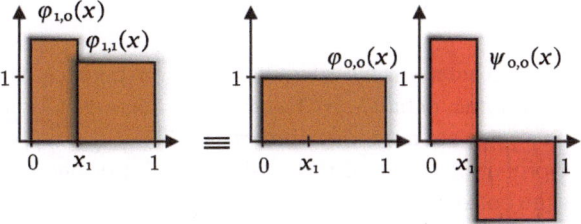

Fig. 4.3 The unbalanced Haar functions family. The scaling functions $\varphi_{1,0}(x)$ and $\varphi_{1,1}(x)$ are defined by a random measurement x_1. The *father wavelet* $\varphi_{0,0}(x) = \varphi(x)$ remains the same as in classic Haar set, but the *mother wavelet* $\psi_{0,0}(x) = \psi(x)$ is unbalanced too

viz., is the sum of the corresponding spacings I_k.[4]

Note that, by design, the unbalanced scaling functions are:

- Compactly supported
- Normalized
- Orthogonal to each other

and therefore are similar to their classic counterparts. However, one should keep in mind the important difference: For a given scale m, the supports χ_{mn} and normalization factors $\sqrt{I_{mn}^{-1}}$ vary randomly for different translation factors n and so do the corresponding scaling functions $\varphi_{mn}(x)$. Therefore, the unbalanced Haar scaling functions no longer possess the *shift invariance* property (w.r.t. to n) of the classic Haar basis functions; see Fig. 4.3.

4.2.1   Linear Approximation

The set $\{\varphi_{Kn}(x)\}$ constitutes the *empirical Haar orthonormal basis* in the interval. Any function $\vartheta(x)$, piecewise-smooth in this interval, can be now approximated by these functions similarly as in the classic case:

$$\vartheta_K(x) = \sum_{n=0}^{2^K-1} \alpha_{Kn}\varphi_{Kn}(x) \tag{4.45}$$

with the approximation coefficients, α_{Kn}, being the appropriate inner products

$$\alpha_{Kn} = \int_{\chi_{Kn}} \vartheta(x)\,\varphi_{Kn}(x)\,dx. \tag{4.46}$$

---

[4]Note that I_{mn} are spacings of order 2^L; cf. [25, Chap. 11.2].

Global (Integrated) Convergence

The following counterparts of the classic Haar convergence and convergence rate lemmas exist for the unbalanced Haar approximation. Observe that all the coefficients $\hat{\alpha}_{Kn}$ in (4.46) are random variables, and hence the approximation errors results hold in a probabilistic sense (viz., with probability one (w.p.1)). The first lemma describes the convergence conditions.

Lemma 4.7. *If the number of measurements N and the scale K tend to infinity so that*

$$K \leq \log_2 N, \tag{4.47}$$

then the integrated squared error of the approximation of $\vartheta (x)$ by $\vartheta_K (x)$ in (4.45) vanishes (w.p.1)

$$\mathrm{ISE}\,\vartheta_K = \int_0^1 [\vartheta_K (x) - \vartheta (x)]^2 \, \mathrm{d}x \to 0 \text{ as } N, K \to \infty. \tag{4.48}$$

Proof. The proof is only sketched: One can follow the reasoning in [147, Chap. 1.2.2] and exploit the fact that $\max_{n=1,\ldots,2^K} \{I_{Kn}\}$ vanishes (w.p.1) with K tending to infinity (see e.g. [57, L. C.11] and [134, Theorem 2.1]). ∎

Convergence Rates

As in the classic case (cf. Lemma 4.4 and e.g., [96, 133, Chap. 9]), also in the unbalanced Haar basis case, the rate of the approximation error decay depends on the smoothness of the target function:

Lemma 4.8. *If the approximated function $\vartheta (x)$ is Lipschitz, then (w.p.1)*

$$\mathrm{ISE}\,\vartheta_K = \mathcal{O}\left(2^{-2K} \log^3 N\right). \tag{4.49}$$

If $\vartheta (x)$ is piecewise-Lipschitz, then (w.p.1)

$$\mathrm{ISE}\,\vartheta_K = \mathcal{O}\left(2^{-K} \log N\right). \tag{4.50}$$

Proof. The proof is rather standard (cf., e.g., [72] and the counterpart in Sect. 4.1.10) with the only modification that, due to the randomness of scaling functions $\varphi_{mn} (x)$, we have for all $n = 0, \ldots, 2^K - 1$ that (cf. [57, L. C.11], [134, Theorem 2.1] and (4.47)):

$$\max_n \{I_{Kn}\} = \mathcal{O}\left(2^{-K} K\right) = \mathcal{O}\left(2^{-K} \log N\right) \tag{4.51}$$

w.p.1, viz., that asymptotically all scaling functions at some scale K have their support lengths bounded by a common quantity in (4.51), which (asymptotically) is only slightly larger than the support length 2^{-K} of the classic Haar functions.

Denote $\operatorname{supp} \varphi_{Kn} = S_{Kn}$ and recall that $\int_{S_{Kn}} dx = I_{Kn}$. We have that

$$\text{ISE}\,\vartheta_K = \int_0^1 \left[\sum_{n=0}^{2^K-1} a_{Kn} \varphi_{Kn}(x) - \vartheta(x) \right]^2 dx \tag{4.52}$$

$$= \sum_{n=0}^{2^K-1} \int_{S_{Kn}} \sum_{n=0}^{2^K-1} [a_{Kn}\varphi_{Kn}(x) - \vartheta(x)]^2 \, dx$$

$$= \sum_{n=0}^{2^K-1} \int_{S_{Kn}} [a_{Kn}\varphi_{Kn}(x) - \vartheta(x)]^2 \, dx,$$

where for the expression in square brackets, we have for any n such that $x \in S_{Kn}$ that

$$a_{Kn}\varphi_{Kn}(x) - \vartheta(x) = \int_{S_{Kn}} \vartheta(v)\varphi_{Kn}(v)\,dv \cdot \varphi_{Kn}(x) - \vartheta(x)$$

$$= \int_{S_{Kn}} \vartheta(v)\varphi_{Kn}(v)\,dv \cdot \varphi_{Kn}(x)$$

$$- \vartheta(x)\,\varphi_{Kn}(x) \underbrace{\int_{S_{Kn}} \varphi_{Kn}(v)\,dv}_{=1 \text{ since } x \in S_{Kn}}$$

$$= \underbrace{\varphi_{Kn}(x)}_{=\sqrt{I_{Kn}^{-1}} \text{ since } x \in S_{Kn}} \int_{S_{Kn}} [\vartheta(v) - \vartheta(x)]\varphi_{Kn}(v)\,dv.$$

For the Lipschitz functions, this leads to the following bound:

$$\sqrt{I_{Kn}^{-1}} \int_{S_{Kn}} |\vartheta(v) - \vartheta(x)|\varphi_{Kn}(v)\,dv \leq c I_{Kn}^{-1} \int_{S_{Kn}} |v - x|\,dv \tag{4.53}$$

$$\leq c I_{Kn}^{-1} \cdot I_{Kn} \int_{S_{Kn}} dv$$

$$= c I_{Kn},$$

and hence

$$\int_{S_{Kn}} [a_{Kn}\varphi_{Kn}(x) - \vartheta(x)]^2 \, dx \leq c \int_{S_{Kn}} I_{Kn}^2 \, dx = c I_{Kn}^3. \tag{4.54}$$

Eventually, for the approximation error ISE ϑ_K expressed in terms of I_{Kn}, it holds that

$$\text{ISE}\,\vartheta_K \leq c \sum_{n=0}^{2^K-1} I_{Kn}^3.$$

Taking now the bound in (4.51) yields that with the probability one

$$\text{ISE}\,\vartheta_K \leq c \sum_{n=0}^{2^K-1} \left(2^{-K} \log_2 N\right)^3 = c2^{-2K} \log_2^3 N.$$

In case of piecewise-Lipschitz functions with q jumps, for the bound in (4.53), it holds that

$$c I_{Kn}^{-1} \int_{S_{Kn}} dv \leq c,$$

for all $n = 0, \ldots, 2^K - 1$. Hence, the integral in (4.54) is bounded by $c I_{Kn}$, and for the resulting approximation error, we eventually have that

$$\text{ISE}\,\vartheta_K \leq c \sum_{n=0}^{2^K-q-1} \left(2^{-K} \log_2 N\right)^3 + q2^{-K} \log_2 N \tag{4.55}$$

$$\leq c2^{-K} \log_2 N. \qquad \blacksquare$$

Clearly, this is the randomness of the unbalanced scaling functions (viz., their support lengths) which implies the presence of the multiplicative factor $K \sim \log N$ in the bound (4.51) and makes the convergence rate of the approximation errors (4.49) and (4.50) slightly slower than in the classic case.

4.2.2 Unbalanced Haar Wavelets

We will now shortly introduce the unbalanced Haar wavelets $\psi_{mn}(x)$. Observe that the crucial properties of the classic Haar wavelets, that is:

- Orthonormality
- Compactness of their support
- A single vanishing moment

will be preserved by the unbalanced ones if they are defined as

$$\psi_{m-1,n}(x) = \sqrt{I_{m-1,n}^{-1}} \sqrt{\frac{I_{m,2n+1}}{I_{m,2n}}} \cdot \chi_{m,2n}(x)$$

$$- \sqrt{I_{m-1,n}^{-1}} \sqrt{\frac{I_{m,2n}}{I_{m,2n+1}}} \cdot \chi_{m,2n+1}(x)$$

$$= \sqrt{\frac{I_{m,2n+1}}{I_{m-1,n}}} \cdot \varphi_{m,2n}(x) - \sqrt{\frac{I_{m,2n}}{I_{m-1,n}}} \cdot \varphi_{m,2n+1}(x), \tag{4.56}$$

in which the scaling function components are weighted so that, for any m, n the following integral is zero

$$\int_0^1 \psi_{mn}(x)\, dx = 0,$$

Due to the presence of the (random) normalization factor $\sqrt{I_{mn}^{-1}}$, the unbalanced wavelets remain equally normalized, i.e.,

$$\int_0^1 \psi_{mn}^2(x)\, dx = 1.$$

Finally, since they are piecewise-constant functions, they are orthogonal to each other as well as the classic Haar wavelets.

We can now, for any given measurements number N (being a dyadic integer), propose the following unbalanced multiresolution approximation (cf. the multiscale form in (4.45)):

$$\vartheta_{MK}(x) = \sum_{n=0}^{2^M-1} \alpha_{Mn}\varphi_{Mn}(x) + \sum_{m=M}^{K-1}\sum_{n=0}^{2^m-1} \beta_{mn}\psi_{mn}(x), \qquad (4.57)$$

where $K \le \log_2 N$ and where the wavelet expansion coefficients are

$$\beta_{mn} = \int_{\chi_{mn}} \vartheta(x)\, \psi_{mn}(x)\, dx.$$

Clearly, $\vartheta_{MK}(x) = \vartheta_K(x)$ by virtue of the wavelet definition formula in (4.56), and hence, the above Lemmas 4.7 and 4.8 hold true for the multiresolution approximation $\vartheta_{MK}(x)$.

4.2.3 Unbalanced Haar Transform

In spite of the fact that the unbalanced Haar functions are no more scaled translations of the pair of their parent functions, the fast algorithm for computing the unbalanced expansion coefficients is still available; cf. [41]. The lifting steps of the unbalanced Haar wavelet transform are of the following form

$$\beta_{m-1,n} = \alpha_{m,2n}\sqrt{\frac{I_{m,2n+1}}{I_{m-1,n}}} - \alpha_{m,2n+1}\sqrt{\frac{I_{m,2n}}{I_{m-1,n}}}, \qquad (4.58)$$

$$\alpha_{m-1,n} = \alpha_{m,2n}\sqrt{\frac{I_{m-1,n}}{I_{m,2n}}} - \beta_{m-1,n}\sqrt{\frac{I_{m,2n+1}}{I_{m,2n}}}.$$

To get the inverse transform step, we only need to perform the above operation backwards, i.e.,

$$\alpha_{m,2n} = \alpha_{m-1,n} \sqrt{\frac{I_{m,2n}}{I_{m-1,n}}} + \beta_{m-1,n} \sqrt{\frac{I_{m,2n+1}}{I_{m-1,n}}}, \tag{4.59}$$

$$\alpha_{m,2n+1} = \alpha_{m,2n} \sqrt{\frac{I_{m,2n+1}}{I_{m,2n}}} - \beta_{m-1,n} \sqrt{\frac{I_{m-1,n}}{I_{m,2n}}}.$$

4.2.4   UHT and EZW Nonlinear Approximation

We examine the nonlinear approximation which, as in the classic Haar case, has the form

$$\vartheta_{MK}(x) = \sum_{n=0}^{2^M-1} \alpha_{Mn} \varphi_{Mn}(x) + \sum_{m=M} \sum_{n \in Q_m}^{K-1} \beta_{mn} \psi_{mn}(x) \tag{4.60}$$

$$= \vartheta_{\mathrm{L}}(x) + \vartheta_{\mathrm{NL}}(x),$$

consisting of the *linear* (*nonadaptive*) multiscale approximation, $\vartheta_{\mathrm{L}}(x)$, and of the *nonlinear* (*adaptive*) part, $\vartheta_{\mathrm{NL}}(x)$, constructed by the EZW approximation scheme, with the wavelet coefficients β_{mn} corresponding to the unbalanced wavelets $\psi_{mn}(x)$ whose supports include jumps.

Convergence and Convergence Rates

The convergence conditions for the nonlinear approximation in (4.60) remain the same as for the linear one in (4.57)—with the scale M taking the role of K (in other words, the convergence is not affected by the presence of the nonlinear part). To establish the convergence rate of the nonlinear approximation based on the EZW scheme, we will exploit the equivalent form of the nonlinear algorithm (4.57), which uses only unbalanced scaling functions. Recall that EZW algorithm creates the "vertical" structures corresponding to the influence cones generated by the jumps of $\vartheta(x)$. The cones consist of stacked wavelet function coefficients associated with wavelet functions which supports at all scales $m = M, \ldots, K-1$ contain the jump points; cf. Fig. 4.2. As a result, we get the approximant which scale varies from M to K and depends on the translation factor n:

$$\vartheta_{MK}(x) = \sum_{k=1}^{N} \alpha_{K(n),n} \varphi_{K(n),n}, \quad K(n) \in \{M, \ldots, K\}. \tag{4.61}$$

The function $\vartheta(x)$ is therefore approximated at the scale 2^K in the q jump intervals of the length of order $\mathcal{O}\left(2^{-(K-1)}(K-1)\right)$ and, then, in the q intervals of the length of order $\mathcal{O}\left(2^{-(K-2)}(K-2)\right)$ at the scale 2^{K-1}. In general, in the q intervals Δ_{K-m-1} of the length of order $\mathcal{O}\left(2^{-(K-m-1)}(K-m-1)\right)$, the approximation scale is 2^{K-m}. Hence, the integrated approximation error ISE ϑ_{MK} can be decomposed as follows (cf. (4.52) and (4.55)):

$$
\text{ISE } \vartheta_{MK} = \int_0^1 \left[\sum_{n=0}^{2^K-1} a_{K(n),n} \varphi_{K(n),n}(x) - \mu(x) \right]^2 dx
$$

$$
\leq cq \sum_{n=0}^{2^M-1} I_{Mn}^{-3} + cq \max_{n=0,\dots,2^K-1} I_{Kn}^{-1}
$$

$$
\leq c2^M q \max_{n=0,\dots,2^M-1} I_{Mn}^{-3} + cq \max_{n=0,\dots,2^K-1} I_{Kn}^{-1},
$$

where the first term is the linear approximation error bound while the other is the approximation error bound of the nonlinear part. By virtue of the bound in (4.51), we have thus obtained the lemma:

Lemma 4.9. *If $\vartheta(x)$ is piecewise-Lipschitz, then the nonlinear approximant in (4.60) for the approximation scale K selected such that*

$$
K = 2M,
$$

approximates $\vartheta(x)$ with the error of order

$$
\text{ISE } \vartheta_{MK} \leq c2^{-2M} \log_2^3 N. \tag{4.62}
$$

Chapter 5
Identification Algorithms

Abstract Four nonparametric Haar regression estimates are proposed and applied to the system nonlinearity identification (recovery). The algorithms are various implementations of the local average paradigm and produce regressograms of the nonlinearity. Three of them are based on the classic Haar series expansion. The first is of a quotient form and resembles the Nadaraya-Watson kernel regression estimate. The second utilizes ordered measurements, while the third maps the measurements using the empirical distribution function. The last one is a version of the second algorithm but exploits the unbalanced Haar series. Two variants of each algorithm, linear and nonlinear, are introduced and studied. The former is based on standard linear Haar approximations. The latter employs the nonlinear (EZW-based) approximation schemes. Convergence conditions and convergence rates of all algorithms are established. The interpolation routine, based on the regressograms (and applicable when the continuous estimate is desired), is eventually derived.

In this chapter we present four Haar wavelet identification algorithms:

- *Quotient orthogonal series* (QOS) algorithm
- *Order statistics* (OS) algorithm
- *Empirical distribution* (ED) algorithm
- *Empirical orthogonal series* (EOS) algorithm

All algorithms use Haar bases; all implement a *local (weighted) averaging paradigm* (cf., e.g., [58, Chap. 2]) and have a common piecewise-constant (*i.e.*, *regressogram*; see [103]) form, but nevertheless, they are derived from various approaches and in distinct ways cope with the randomness of the input–output signals:

1. In the *QOS* algorithm, the empirical expansion coefficients are computed as simple local averages of the output measurements, y_k's.
2. In the *order statistics* algorithm, these outputs are weighted by the random distances (spacings) between the adjacent-ordered input measurements.

P. Śliwiński, *Nonlinear System Identification by Haar Wavelets*, Lecture Notes in Statistics 210, DOI 10.1007/978-3-642-29396-2_5,
© Springer-Verlag Berlin Heidelberg 2013

3. In the *empirical distribution* algorithm, the empirical coefficients are again simple local averages (as in QOS algorithm) since the randomly scattered input data are—prior to averaging—mapped onto to the equidistant grid by their empirical distribution function.
4. In the *EOS* algorithm, the coefficients are averages of output measurements y_k weighted by the corresponding spacings as in OS algorithm; however, the orthogonal basis is unknown a priori and design-adapted, that is, generated from the input data collected during the identification experiment; cf. Sect. 4.2.

In attempt to exploit the localization properties of wavelet expansion and to improve the convergence rates of the identification algorithms in case when the identified nonlinearities are discontinuous (piecewise-Lipschitz), we propose the nonlinear approximation-based variants of the above algorithms. They utilize a common *heuristic* EZW-like nonlinear approximation scheme and are founded on the following observations (see sections "EZW Approximation Scheme" and 4.2.4):

• The order of the wavelet expansion coefficients decays slower (as their scale grows) in regions where the nonlinearity is discontinuous
• Furthermore, these coefficients are localized around the jump points (within the *cones of influence* of these points; see Fig. 4.2)
• All expansion coefficient estimates (referred further to as *empirical coefficients*) have variances of the same order (cf. e.g., [62, 63])[1]

The conditions for the global (integrated) mean square convergence and asymptotic convergence rates are presented for each algorithm. In all cases—if not explicitly stated—Assumptions 1–4 about the underlying systems from the Chap. 2 are in force. We show, in particular, that all linear algorithms converge to the true system nonlinearity $\mu(x)$[2] with the optimal (or near optimal[3]) convergence rates for Lipschitz nonlinearities. For nonlinear algorithms, we establish both the best and the worst-case convergence rates.

Remark 5.1. The EZW scheme presented in section "EZW Approximation Scheme" assumes (for the simplicity of its presentation) that the number of discontinuities is known. Since it is rather unrealistic assumption (yet considered in the literature, cf., e.g., [63, Sect. 2.5.3]), we relax it and in all nonlinear identification

---

[1]In nonparametric regression estimation the estimates of the orthogonal wavelet coefficients have the same variance error order $\mathcal{O}(1/N)$, regardless of the scale and translation factors. This fact is well known in the statistical wavelet literature and commonly used in nonlinear algorithms constructions (cf., e.g., [63]). Our goal is focused on showing that this property holds (and can be exploited) in the proposed algorithms under the specific conditions imposed by the nonparametric system identification assumptions.

[2]In the algorithm description, we will rather use x and x_k to denote the algorithm inputs (instead of the u and u_k previously used to describe the inputs of the identified systems).

[3]In case of the EOS algorithm, the rate is actually log-optimal (i.e., slowered by the $\log N$ factor).

algorithms, we allow the number of jumps in the nonlinearity and in the input probability density function to be unknown.

5.1 Numerical Illustrations

The formal examinations of the algorithms' asymptotic properties are intertwined with several numerical experiments illustrating their behavior for small and moderate measurement sets.[4] The test Hammerstein systems (common for all algorithms) consist of an input nonlinearity being either a piecewise-constant function

$$m(u) = \frac{1}{2} \left\lfloor \frac{1}{2} + 5\left(u - \frac{1}{2}\right) \right\rfloor, \tag{5.1}$$

or a piecewise-polynomial one

$$m(u) = -10\left(2u^3 - 3u^2 + u\right) - \frac{1}{5}\operatorname{sgn}\left(u - \frac{1}{2}\right), \tag{5.2}$$

where $\operatorname{sgn}(u) = u/|u|$ (with $\operatorname{sgn}(0) = 0$) denotes the standard signum function.

The linear dynamic element accompanying the identified nonlinear part has one of the following infinite length (and oscillating) impulse responses[5]

$$\lambda_i = \lambda^{-i} \text{ with } \lambda = 0, -\frac{1}{4}, -\frac{1}{2} \text{ or } -\frac{3}{4}. \tag{5.3}$$

The systems input is driven by a random *i.i.d.* signal with a triangular

$$f(u) = 2\left(1 - |2u - 1|\right) \cdot \chi_{[0,1]}(u), \tag{5.4}$$

or a piecewise-constant probability density function

$$f(u) = 9 \sum_{i=1}^{9} f_i \cdot \chi_{\left[\frac{i-1}{9}, \frac{i}{9}\right)}(u), \tag{5.5}$$

with $f_i = \{0.1, 0.2, 0.02, 0.05, 0.26, 0.05, 0.02, 0.2, 0.1\}$. The system outputs are disturbed by a white zero-mean uniformly distributed noise $z_k \sim U[-0.1, 0.1]$.

Remark 5.2. Note that all test functions (5.1), (5.2), (5.4), and (5.5) have infinite representations in the Haar wavelet basis (including the piecewise-constant

---

[4]The C++ implementations of the algorithms are available for download from the author's site http://diuna.iiar.pwr.wroc.pl/sliwinski/software/Book.zip.

[5]The case $\lambda = 0$ corresponds to the memoryless (static) nonlinear system. Taking other λs makes the experimental systems dynamic and results in $SNR = 3, 1$ and $1/3$, respectively; cf. 3.2 and Fig. 3.1.

characteristic in (5.1) and the density function in (5.5) which both have jumps located in nondyadic points).

The experimental outcomes are displayed with the help of two types of diagrams. The first demonstrates a single experiment run performance of an estimate for the systems with impulse responses as in (5.3) and for sets of $N = 4, 8, \ldots, 512$ measurement pairs. The other, for a given input *pdf* function $f(u)$ and the dynamic subsystem with $\lambda = -1/4$ (viz., for $SNR = 3$; cf. 3.2), shows the shapes of the tested nonlinear characteristic $m(u)$ and of the density function $f(u)$ against the estimate realization $\hat{\mu}_K(u)$ computed for $N = 512$; see Figs. 5.2–5.13.

5.2   Quotient Orthogonal Series Algorithm

Let $\{(x_k, y_k)\}$ be the measurements set (i.e., let the algorithms inputs x_k be just the system inputs u_k). The *QOS algorithm* has the form

$$\hat{\mu}_K(x) = \frac{\displaystyle\sum_{n=0}^{2^K-1} \hat{\alpha}_{Kn} \varphi_{Kn}(x)}{\displaystyle\sum_{n=0}^{2^K-1} \hat{a}_{Kn} \varphi_{Kn}(x)}, \tag{5.6}$$

where the empirical expansion coefficients $\hat{\alpha}_{Mn}$ and $\hat{a}_{Mn}$ are the empirical means[6]:

$$\begin{bmatrix} \hat{\alpha}_{Mn} \\ \hat{a}_{Mn} \end{bmatrix} = \frac{1}{N} \sum_{k=1}^{N} \begin{bmatrix} y_k \\ 1 \end{bmatrix} \varphi_{Mn}(x_k). \tag{5.7}$$

Because of compactness of the supports of the Haar scaling functions and because their supports do not overlap, the algorithm can be written in a simpler form

$$\hat{\mu}_K(x) = \left. \frac{\hat{\alpha}_{Kn}}{\hat{a}_{Kn}} \right|_{n = \lfloor 2^K x \rfloor}, \tag{5.8}$$

since for each x, only single coefficients from the numerator and denominator are active and determine the algorithm output; cf. the implementation in Sect. 6.1.1. We can now easily recognize that the QOS algorithm simply computes local averages of the nonlinearity $\mu(x)$ in each interval $\mathrm{supp}\, \varphi_{Kn} = \left[2^{-K} n, 2^{-K}(n+1) \right)$, using a random number N_{Kn} of those output measurements y_k, which corresponding inputs x_k fall into this interval (see Fig. 5.1):

---

[6]The formulas for the empirical coefficients in the algorithm numerator and denominator are presented in a joint compact matrix-like form to emphasize their close similarity.

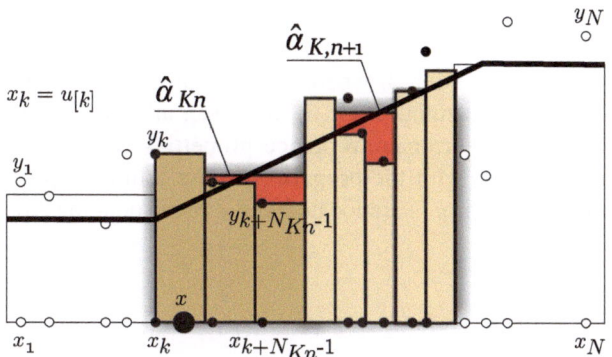

Fig. 5.1 The illustration of the idea behind the QOS algorithm. Note that the number of measurements taken into account is random and each measurement is weighted by the same factor $1/N_{Kn}$

$$\hat{\mu}_K(x) = \frac{1}{N_{Kn}} \sum_{\{k : x_k \in \mathrm{supp}\, \varphi_{Kn}\}} y_k. \tag{5.9}$$

Remark 5.3. The equivalent kernel-like representation of the QOS algorithm can be derived from (5.7) and (5.8):

$$\hat{\mu}_K(x) = \frac{\sum\limits_{k=1}^{N} y_k \vartheta_K(x, x_k)}{\sum\limits_{k=1}^{N} \vartheta_K(x, x_k)}, \tag{5.10}$$

observing that, at the scale K, the Haar reproducing kernel is of the form

$$\vartheta_K(x, v) = \varphi_{Kn}(x)\, \varphi_{Kn}(v) = 2^K \chi_{[0,1)} \left(2^K x - \lfloor 2^K v \rfloor\right)$$

with $n = \lfloor 2^K x \rfloor$ (cf. (4.12)). The form (5.10) naturally resembles the classic Nadaraya-Watson regression function estimate with the simple window (rectangular) kernel $K(x) = \chi_{[0,1)}(x)$:

$$\hat{\mu}_K(x) = \frac{\sum\limits_{k=1}^{N} y_k K\left(\frac{x - x_k}{h}\right)}{\sum\limits_{k=1}^{N} K\left(\frac{x - x_k}{h}\right)},$$

where h is the so-called *bandwidth parameter*; cf. [106, 149] and [58, 65, Chap. 5].

In the following theorems we characterize the convergence properties of the QOS algorithm (5.6) assuming that *the output noise,* z_k, *is bounded*. This is a slight departure from the Assumption 4, where it is assumed that its variance, var z_k, is bounded but not the noise signal itself. It, however, allows to examine the global (integrated) mean square error convergence properties of the estimate and then to compare them directly with the properties of the estimates produced by other proposed algorithms. All proofs are located in the corresponding Appendix A.1.

5.2.1 Convergence

In the theorem below we present the simple conditions for the algorithm to converge globally in the MISE error sense, defined in the standard way

$$\text{MISE}\, \hat{\mu}_K = E \int_0^1 [\hat{\mu}_K(x) - \mu(x)]^2 \, \mathrm{d}x. \tag{5.11}$$

Theorem 5.1. *If the scale K of the estimate in (5.6) is selected so that*

$$K \to \infty \text{ and } 2^K / N \to 0 \text{ as } N \to \infty,$$

then the QOS algorithm converges to the nonlinearity $\mu(x)$ globally with growing number of measurements N, that is, the MISE error vanishes with the measurements number N growing large:

$$\text{MISE}\, \hat{\mu}_K \to 0 \text{ as } N \to \infty.$$

In a view of Assumptions 1–3, the theorem shows that with the increasing number of measurements N and for the estimate scale K selected (for instance) so that $K = \eta \log_2 N$, any $\eta \in (0, 1)$, the algorithm converges globally to any piecewise-continuous nonlinearity for:

- Any piecewise-continuous input probability density function
- Any asymptotically stable system dynamics
- Any bounded noise (correlated or white)

5.2.2 Convergence Rates

Once we have established the convergence, we can characterize the convergence rates of the QOS algorithm. In the following theorems, we derive the rates for Lipschitz and piecewise-Lipschitz nonlinearities and input probability densities.

Theorem 5.2. *Let the nonlinearity $\mu(x)$ and the input probability density function $f(x)$ be Lipschitz. If the scale K of the estimate in (5.6) is selected using the formula[7]*

$$K = \tfrac{1}{3} \log_2 N, \tag{5.12}$$

then the QOS algorithm converges to this nonlinearity globally, in the MISE error sense, with the rate

$$\text{MISE } \hat{\mu}_K = \mathcal{O}\left(N^{-2/3}\right). \tag{5.13}$$

Recall that the convergence rate $\mathcal{O}\left(N^{-2/3}\right)$ is known to be *optimal* (the fastest possible) for such a class of nonlinearities; cf. [138]. The next theorem establishes the convergence rate of the QOS algorithm when either or both $\mu(x)$ and $f(x)$ are piecewise-Lipschitz (discontinuous) functions.

Theorem 5.3. *Let the nonlinearity $\mu(x)$ or the input probability density function $f(x)$ be piecewise-Lipschitz. If the scale K of the estimate in (5.6) is selected using the formula*

$$K = \tfrac{1}{2} \log_2 N, \tag{5.14}$$

then the QOS algorithm converges to this nonlinearity globally (in the MISE error sense) with the rate

$$\text{MISE } \hat{\mu}_K = \mathcal{O}\left(N^{-1/2}\right). \tag{5.15}$$

The theorem shows that in the presence of discontinuities the convergence rate is of order $\mathcal{O}\left(N^{-1/2}\right)$, that is, becomes slower than for uniformly Lipschitz functions. Observe also that this rate is also optimal for such discontinuous functions and is not attained by the classic (trigonometric or polynomial series) orthogonal series estimates; see [29] and cf. [57, Chap. 6]. Note that in practice we rarely know in advance whether the nonlinearity is continuous or not. The above theorem assures however that the slower rate in (5.15) is guaranteed in either case. The next corollary describes, in turn, the consequences of a wrong assumption that the nonlinearity is Lipschitz (when in fact it possesses jumps).

Corollary 5.1. *If the nonlinearity (or the density function) is piecewise-Lipschitz but the scale selection rule (5.12) is used, then the algorithm still converges, but the convergence rate is slowed down to $\mathcal{O}\left(N^{-1/3}\right)$.*

The final theorem reveals that the Lipschitz convergence rate $\mathcal{O}\left(N^{-2/3}\right)$ can nevertheless be preserved for some discontinuous piecewise-Lipschitz nonlinearities (and/or density functions); cf. Corollary 4.1 and Remark 4.6.

Theorem 5.4. *If the nonlinearity $\mu(x)$ or the density function $f(x)$ (or both) is discontinuous but has jumps only at* dyadic *(binary rational) points, then for*

---

[7]Since K is an integer, one should quantize the formula $1/3 \cdot \log_2 N$ using, e.g., either *ceiling* $\lceil \cdot \rceil$ or *floor* $\lfloor \cdot \rfloor$ functions (cf., e.g., [44]). It however does not influence the asymptotic properties considered in the chapter.

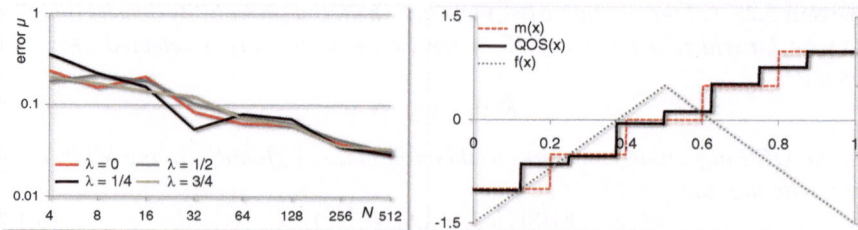

Fig. 5.2 The *quotient orthogonal series* (QOS) estimate is robust against the dynamics (*left*) but rather poorly localizes jumps at nondyadic points (*right*); cf. Fig. 3.1

the scale selection rule (5.12), the algorithm converges globally with the rate $\mathcal{O}\left(N^{-2/3}\right).$

That the estimate is "blind" to the presence of jumps located in binary rational points is an advantageous consequence of the construction of the classical Haar functions which, being dyadic translations of the father and mother wavelet functions, have jumps at these points as well; cf. Sect. 4.1.

5.3 Nonlinear QOS Algorithm

In attempt to improve the convergence rate of the identification algorithm in the more general case when discontinuous characteristic and/or input density function have jumps in arbitrary points (e.g., at a nondyadic point $x = 0.1$), we propose the nonlinear counterpart of the above QOS algorithm which implements the EZW scheme presented in Sect. 4.1.13.

The nonlinear quotient orthogonal series algorithm (NQOS) employs both Haar scaling and wavelet functions, and is of the following form, cf. (5.6):

$$\hat{\mu}_{MK}\left(x\right) = \frac{\displaystyle\sum_{n=0}^{2^M-1} \hat{\alpha}_{Mn}\varphi_{Mn}\left(x\right) + \sum_{m=M}^{K-1}\sum_{n\in Q_m} \hat{\beta}_{mn}\psi_{mn}\left(x\right)}{\displaystyle\sum_{n=0}^{2^M-1} \hat{a}_{Mn}\varphi_{Mn}\left(x\right) + \sum_{m=M}^{K-1}\sum_{n\in Q_m} \hat{b}_{mn}\psi_{mn}\left(x\right)}, \qquad (5.16)$$

where the empirical coefficients are given similarly as in the linear variant (5.7):

$$\begin{bmatrix} \hat{\alpha}_{Mn} & \hat{\beta}_{mn} \\ \hat{a}_{Mn} & \hat{b}_{mn} \end{bmatrix} = \frac{1}{N}\sum_{k=1}^{N} \begin{bmatrix} y_k & 0 \\ 1 & 0 \end{bmatrix} \begin{bmatrix} \varphi_{Mn}\left(x_k\right) & \psi_{mn}\left(x_k\right) \\ \psi_{mn}\left(x_k\right) & \varphi_{Mn}\left(x_k\right) \end{bmatrix}, \qquad (5.17)$$

and where Q_m are sets containing, at each scale m, the q_M translation indices n of the empirical wavelet coefficients $\hat{\beta}_{mn}$, determined by the EZW algorithm as being

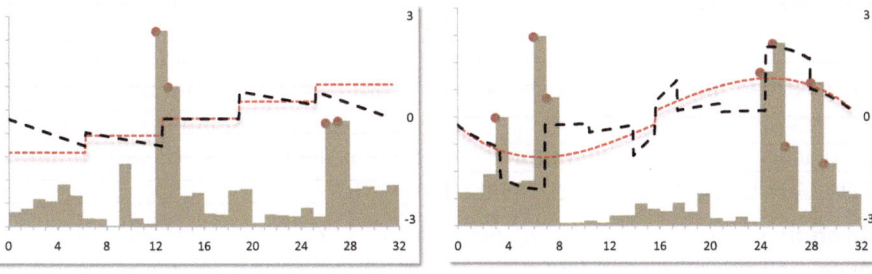

Fig. 5.3 In the NQOS algorithm, the EZW scheme detects discontinuities generated by both the nonlinearity and the density function. All empirical cones are displayed against the nonlinearities, (5.1) and (5.2), and input densities (5.4) and (5.5), on the *left* and *right* diagram, respectively

located in the cones of influence. The number q_M will be referred to as the empirical cone number. The implementation of the EZW is based on *the heuristic observation* that, likewise the genuine coefficients β_{mn}, the large empirical coefficients $\hat{\beta}_{mn}$ are clustered in the vicinity of the jump points of $\mu(x)$ or $f(x)$ (and is given in detail in Sect. 6.3).[8] In result, the numerator of the estimate consists of the linear part comprising the empirical scaling function coefficients $\hat{\alpha}_{Mn}$ evaluated at the scale M and of the adaptive nonlinear part which includes $q_M(K-M)$ empirical wavelet coefficients $\hat{\beta}_{mn}$ (that is, q_M coefficients per each scale $m = M, \ldots, K-1$) forming the empirical influence cones; see Fig. 5.3. The empirical wavelet coefficients $\hat{b}_{mn}$ in the denominator are not selected in a separate run of the EZW routine but are simply chosen to match those in the numerator. In effect, the translation factors n of the coefficients $\hat{b}_{mn}$ are taken from the same sets Q_m.

Remark 5.4. The numerator of the quotient algorithms estimates in fact a product of the nonlinearity and of the input probability density function, $g(x) = \mu(x) f(x)$. Thus, in particular, the locations of jumps of $f(x)$ remain in $g(x)$ in the same points (along with the jumps induced by the nonlinearity $\mu(x)$); see Appendix A.1 for details. Subsequently, if the locations of the empirical coefficients $\hat{\beta}_{mn}$ selected by the EZW algorithm are valid for the numerator, they are also valid for the denominator.

---

[8] We emphasize that the *heuristic implementation* of the EZW nonlinear approximation scheme does not take into account the fact that the empirical wavelet coefficients are heavily disturbed by the random (system and external) noises and, subsequently, that the values of the empirical coefficients $\hat{\beta}_{mn}$ need not to precisely indicate the locations of the actual jumps and the corresponding cones of influence in the nonlinearity. Nevertheless, since each empirical cone is selected according to the values of $K - M$ empirical coefficients rather than according to the value of the single one (as it would occur in a *naive* implementation of the N-term nonlinear approximation scheme where isolated coefficients would have been included into the nonlinear part), the EZW-based approach seems to be more robust against the noise.

5.3.1   Convergence

The next two theorems establish the convergence conditions and the rates of the convergence for the *worst case* (e.g., for the most pessimistic case when all wavelet coefficients $\hat{\beta}_{mn}$ are selected incorrectly) and for the *best case* (i.e., when all the cones of influence are properly detected), respectively.

Theorem 5.5. *If the scale M of the estimate in (5.16) and the empirical cone number q_M are selected so that*

$$\begin{cases} M \to \infty \\ q_M \to \infty \end{cases} \text{ and } \quad \begin{array}{l} \frac{2^M}{N} \to 0 \\ \frac{q_M M}{2^M} \to 0 \end{array} \quad , \text{ as } N \to \infty, \tag{5.18}$$

and moreover the scale K is selected as

$$K = \nu M, \tag{5.19}$$

for any $\nu \geq 0$, then the nonlinear QOS algorithm converges to the nonlinearity $\mu(x)$ globally, in the MISE error sense, with a growing number of measurements, i.e., it holds that

$$\text{MISE } \hat{\mu}_{MK} \to 0 \text{ as } N \to \infty.$$

The theorem says that the convergence conditions of the nonlinear algorithm are basically the same as for the linear one. That is, application of the EZW nonlinear algorithm does not actually interfere with the convergence provided that the rather weak restrictions imposed on q_M and K are satisfied. The restrictions in (5.18) imposed on q_M assure that the number of the empirical cones created in the algorithm will eventually exceed any arbitrarily large (but finite) number of jumps of the identified nonlinearity and the input probability density function. Simultaneously, the rate of growth of the scale factor K selected according to any rule compatible with (5.19) prevents the overall number of empirical wavelet coefficients $q_M(K - M)$ to eventually exceed the number of empirical scaling function coefficients.

5.3.2   Convergence Rates

The other theorem provides with the convergence rates attainable by the algorithm.

Theorem 5.6. *If the linear part scale M of the estimate in (5.16) is selected using the formula*

$$M = \tfrac{1}{3} \log_2 N, \tag{5.20}$$

the scale K of nonlinear part equals

$$K = 2M,$$

and the number of empirical cones is

$$q_M = \log_2 M, \tag{5.21}$$

then the nonlinear QOS algorithm (5.16)–(5.17) converges to the nonlinearity globally, in the MISE error sense, no faster than

$$\text{MISE } \hat{\mu}_{MK} = \mathcal{O}\left(N^{-2/3}\right),$$

but no slower than

$$\text{MISE } \hat{\mu}_{MK} = \mathcal{O}\left(N^{-1/3}\right).$$

Observe that the best-case rate of the nonlinear algorithm (5.16) is the same as the rate of the linear algorithm (5.6) when $\mu(x)$ and $f(x)$ are Lipschitz (or have jumps at binary points, cf. Theorem 5.4). That is, if the empirical wavelet coefficients are properly selected, then the presence of arbitrarily located discontinuities does not deteriorate the nonlinear algorithm convergence rate. Moreover, if the nonlinearity and the density function have no jumps, the convergence rate remains the same as for the linear algorithm, i.e., the best possible $\mathcal{O}\left(N^{-2/3}\right)$.

Remark 5.5. The worst-case convergence rate $\mathcal{O}\left(N^{-1/3}\right)$ of the proposed nonlinear algorithm is equal to the rate obtained by the linear algorithm with the same scale selection rule $K = 1/3 \log_2 N$ but slower than the rate $\mathcal{O}\left(N^{-1/2}\right)$ of the linear algorithm with $K = 1/2 \log_2 N$; cf. (5.13) and (5.15), respectively.

As the final note, observe that all the established convergence rates for the QOS-type algorithms are independent of:

- The type of system dynamics
- The correlation structure of the external (bounded) noise

Nevertheless, the performance of the linear and nonlinear QOS algorithms depends on the smoothness of both the nonlinearity $\mu(x)$ and the probability density function $f(x)$.

Selection of the Empirical Cone Number q_M

The conditions for proper selection of q_M in (5.18) are rather weak as they allow the number of empirical cones to grow with the scale M as a polynomial (in M) of any degree. Therefore, we set q_M as in (5.21) quite arbitrarily (only taking into account the computational aspects; cf. Sect. 6.3—for this selection rule the number of empirical cones grows at the slow (iterated logarithm-like) rate $\mathcal{O}(\log \log N)$; cf. (5.20).

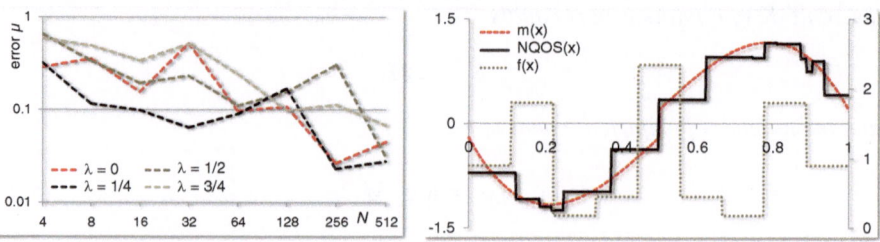

Fig. 5.4 The *nonlinear quotient orthogonal series* (NQOS) estimate is strongly affected by the dynamics (*left*) but quite well restores the shape of the nolinearity (*right*); cf. Fig. 3.1

5.4 Order Statistics (OS) Algorithm

The QOS algorithms identify the nonlinearity $\mu(x)$ indirectly, using the raw set of measurement data $\{(u_k, y_k)\}$ to estimate the quotient of the product $g(x) = \mu(x) f(x)$ and the input probability density function $f(x)$; cf. Remark 5.4 and Appendix A.1. In the following algorithm, the nonlinearity is estimated in a direct way; however, prior to the identification routine, the measurement set $\{(u_k, y_k)\}$, $k = 1, \ldots, N$, is sorted pairwise w.r.t. the increasing input values u_k. The resulting set of ordered pairs is further denoted by $\{(x_k, y_k)\}$, $k = 1, \ldots, N$, and supplemented by the (artifact) boundary measurements $(x_0, y_0) = (0, 0)$ and $(x_{N+1}, y_{N+1}) = (1, 0)$.

The algorithm, referred to as the *order statistics algorithm* (OS), takes the simple form[9]

$$\hat{\mu}_K(x) = \sum_{n=0}^{2^K - 1} \hat{\alpha}_{Kn} \varphi_{Kn}(x), \tag{5.22}$$

where the empirical coefficients,

$$\hat{\alpha}_{Kn} = \sum_{k=1}^{N} y_k \int_{x_{k-1}}^{x_k} \varphi_{Kn}(x)\, dx, \tag{5.23}$$

estimate the scaling function coefficients of $\mu(x)$ in the approximation space V_K (that is, the integrals corresponding to the inner products of the scaling functions $\varphi_{Kn}(x)$ and the recovered nonlinearity $\mu(x)$; see Fig. 5.5 and cf. (4.10)):

$$\alpha_{Kn} = \int_0^1 \mu(u)\, \varphi_{Kn}(x)\, dx = \sum_{k=1}^{N+1} \int_{x_{k-1}}^{x_k} \mu(x)\, \varphi_{Kn}(x)\, dx, \tag{5.24}$$

---

[9]The name of the algorithm is derived from the fact that the sorted input sequence $\{x_k\}$ becomes the *order statistics* of the original input measurements (cf., e.g., [25, 57, App. C.4]).

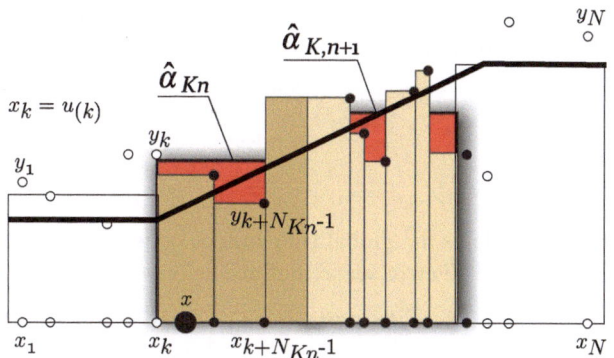

Fig. 5.5 The illustration of the idea behind the OS algorithm. The empirical coefficients are averages of the output measurements, y_ks, weighted by the distance between the spacings $x_k - x_{k-1}$ generated by the corresponding inputs

In practice, to avoid integration, we can—by virtue of the *fundamental theorem of calculus* (see, e.g., [1, Th. 5.3])—use the equivalent integration-free version of the empirical coefficients

$$\hat{\alpha}_{Kn} = \sum_{k=1}^{N} y_k \left[\Phi_{Kn}(x_k) - \Phi_{Kn}(x_{k-1}) \right], \tag{5.25}$$

where $\Phi_{Kn} = \sqrt{2^{-K}} \, \Phi\left(2^K x - n\right)$ are the *indefinite integrals* of the scaling functions φ_{Kn}, i.e., the scaled and translated versions of the indefinite integral of the father wavelet $\varphi(x)$, cf. [129, 133],

$$\Phi(x) = x \cdot \chi_{[0,1)}(x) + \chi_{[1,\infty)}(x).$$

Remark 5.6. One can consider a yet another integration-free version of the empirical coefficients in (5.23), derived directly from the Riemann definition (see, e.g., [119, Chap. 6])

$$\check{\alpha}_{Kn} = \sum_{k=1}^{N} y_k \varphi_{Kn}(x_k)(x_k - x_{k-1}) = \sum_{k=1}^{N} y_k \int_{x_{k-1}}^{x_k} \varphi_{Kn}(x_k) \, \mathrm{d}x.$$

It has been shown however that this formula may lead to worse performance of the estimates since the knowledge of the scaling function (its exact integral) is not exploited (see e.g., [57, Chap. 7]).

We will present the limit properties of the OS algorithm utilizing the pattern similar to the one used for the QOS algorithm: We first establish the convergence conditions and then the convergence rates for Lipschitz and piecewise-Lipschitz nonlinearities, for both linear and nonlinear algorithm variants (all the corresponding proofs are in Appendix A.2).

5.4.1 Convergence

The first theorem formulates the global (integrated) convergence conditions of the linear algorithm in the MISE error sense (defined as previously in (5.11)).

Theorem 5.7. *If the scale K of the estimate in (5.22) is selected so that*

$$K \to \infty \text{ and } 2^K/N \to 0 \text{ as } N \to \infty, \tag{5.26}$$

then the order statistics algorithm converges to the nonlinearity $\mu(u)$ globally, in the MISE error sense, with growing number of measurements, i.e., we have

$$\text{MISE } \hat{\mu}_K \to 0 \text{ as } N \to \infty.$$

Observing that the 2^K factor in the (5.26) is just a number of empirical coefficients in the OS estimate (5.22), the convergence conditions are rather plain: To make the algorithm converge with the growing number of measurements N, it suffices to increase the scale K at any rate so that the number of these coefficients (2^K) is of order $o(N)$. Furthermore, in a view of Assumptions 1–4, the theorem shows that the algorithm converges globally to any piecewise-Lipschitz nonlinearity, for:

- Any piecewise-continuous input probability density function
- Any asymptotically stable system dynamics
- Any second-order noise (correlated or white)

The first two properties are shared with the QOS algorithm. Additionally, the current algorithm converges globally also in presence of the unbounded external noise signals.

5.4.2 Convergence Rates

In the following theorems, the asymptotic convergence rates of the linear OS algorithm are described for different types of nonlinearities.

Theorem 5.8. *Let the nonlinearity $\mu(u)$ be Lipschitz. If the scale K of the estimate in (5.22) is selected using the formula*

$$K = \tfrac{1}{3} \log_2 N, \tag{5.27}$$

then the order statistics algorithm converges to the nonlinearity globally, in the MISE error sense, with the rate

$$\text{MISE } \hat{\mu}_K = \mathcal{O}\left(N^{-2/3}\right).$$

The theorem says that the convergence rate of the OS algorithm (optimal for such class of nonlinearities, cf. [138]) is independent of:

- The type of system dynamics
- The noise correlation structure
- The (lack of) smoothness of the input probability density function

The following theorem establishes the convergence rate of the OS algorithm when $\mu(u)$ is discontinuous.

Theorem 5.9. *Let the nonlinearity $\mu(u)$ be piecewise-Lipschitz. If the scale K of the estimate in (5.22) is selected using the formula*

$$K = \tfrac{1}{2}\log_2 N, \tag{5.28}$$

then the order statistics algorithm converges globally to this nonlinearity in the MISE error sense with the rate

$$\text{MISE } \hat{\mu}_K = \mathcal{O}\left(N^{-1/2}\right).$$

As in the QOS algorithm case, the guaranteed convergence rate decelerates (in general) in the presence of discontinuities. Furthermore, taking an improper scale selection rule has similar consequences as well; cf. Corollary 5.1:

Corollary 5.2. *If the nonlinearity is piecewise-Lipschitz but the scale selection rule in (5.27) is used instead of that in (5.28), then the algorithm convergence rate decelerates further to the rate $\mathcal{O}\left(N^{-1/3}\right)$.*

To complete description of the linear OS algorithm convergence properties, we inspect the case when the nonlinearity is dyadic piecewise-Lipschitz; cf. Corollary 4.1, Remark 4.6, and the analogue Theorem 5.4 (Fig. 5.6).

Theorem 5.10. *If the nonlinearity $\mu(u)$ is discontinuous with the jumps located only in dyadic (binary rational) points, then for the scale selection rule (5.12), the algorithm converges globally with the rate $\mathcal{O}\left(N^{-2/3}\right)$.*

The theorem says that the convergence rate optimal for Lipschitz nonlinearities is maintained for the piecewise-Lipschitz ones if only the jump points are dyadic. Note that—in contrast to the QOS algorithm—this rate remains unaffected even if the input density functions have arbitrarily located jumps.

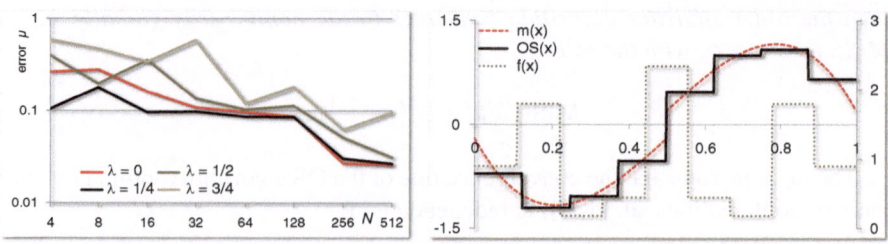

Fig. 5.6 The *order statistics* (OS) estimate perform well against the dynamics (*left*), but less effectively restores the shape of the nonlinearity (*right*); cf. Fig. 5.2

5.5 Nonlinear OS Algorithm

The nonlinear version of the *order statistics* algorithm (NOS) produces the estimate consisting of the linear part with the empirical scaling function coefficients and of the nonlinear one, with the empirical wavelet coefficients selected with the help of EZW scheme (cf. section "EZW Approximation Scheme")

$$\hat{\mu}_{MK}(x) = \sum_{n=0}^{2^M-1} \hat{\alpha}_{Mn}\varphi_{Mn}(x) + \sum_{m=M}^{K-1} \sum_{n\in Q_m} \hat{\beta}_{mn}\psi_{mn}(x), \qquad (5.29)$$

with the empirical coefficients computed as[10]

$$\hat{\alpha}_{Mn} = \sum_{k=1}^{N} y_k \int_{x_{k-1}}^{x_k} \varphi_{Mn}(x)\,\mathrm{d}x \text{ and } \hat{\beta}_{mn} = \sum_{k=1}^{N} y_k \int_{x_{k-1}}^{x_k} \psi_{mn}(x)\,\mathrm{d}x.$$

Each set Q_m, being a subset of the set Q_{MK} of all wavelet empirical coefficients, contains q_M translation indices corresponding to the empirical coefficients $\hat{\beta}_{mn}$ at the scales $m = M, \ldots, K - 1$, which are created by the EZW algorithm and are (tentatively) located in the cones of influences of the jumps; see Sect. 6.3 for implementation of the EZW.

5.5.1 Convergence and Convergence Rates

The following pair of theorems describes the algorithm convergence conditions and the best- and worst-case convergence rates. The first is the counterpart of Theorem 5.5.

---

[10]The equivalent integration-free versions of these coefficients are easy to derive from (5.25).

Theorem 5.11. *If the scale M of the estimate in (5.29) and the empirical cone number q_M are selected so that*

$$\begin{cases} M \to \infty \\ q_M \to \infty \end{cases} \text{ and } \begin{matrix} \frac{2^M}{N} \to 0 \\ \frac{q_M M}{2^M} \to 0 \end{matrix} \quad , \text{ as } N \to \infty, \tag{5.30}$$

and the scale K is selected as

$$K = \nu M,$$

for any $\nu \geq 0$, then the nonlinear orthogonal series algorithm converges to the nonlinearity $\mu(x)$ globally, in the MISE error sense, with the growing number of measurements N, i.e., it holds that

$$\text{MISE } \hat{\mu}_{MK} \to 0 \text{ as } N \to \infty.$$

The theorem says that the convergence conditions of the nonlinear algorithm are virtually the same as for the linear one. That is, the application of the EZW nonlinear algorithm is "asymptotically safe" and does not make the algorithm diverge. The only extra requirements imposed on q_M and K assure that the number of empirical cones created in the algorithm will eventually exceed any arbitrarily finite number of jumps of the identified nonlinearity and, simultaneously, that the size of the nonlinear part (namely the number of wavelet empirical coefficients $\hat{\beta}_{mn}$) will be kept smaller than the size of the linear part.

By virtue of the next theorem, the convergence rate $\mathcal{O}\left(N^{-2/3}\right)$ can be attained by the nonlinear OS algorithm even if the piecewise-Lipschitz nonlinearity has jumps in arbitrary points (cf. Theorem 5.6 and Corollary 5.2).

Theorem 5.12. *Let the nonlinearity $\mu(u)$ be piecewise-Lipschitz. If the scales M and K of the estimate in (5.29) are selected using the formulas*

$$M = \tfrac{1}{3} \log_2 N \text{ and } K = 2M,$$

and the number of empirical cones is

$$q_M = \log_2 M, \tag{5.31}$$

then the nonlinear order statistics algorithm converges to this nonlinearity globally, in the MISE error sense, with the best-case rate

$$\text{MISE } \hat{\mu}_{MK} = \mathcal{O}\left(N^{-2/3}\right),$$

or with the worst-case rate

$$\text{MISE } \hat{\mu}_{MK} = \mathcal{O}\left(N^{-1/3}\right).$$

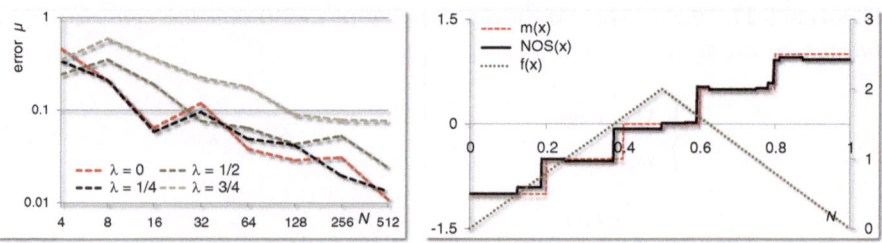

Fig. 5.7 The *nonlinear order statistics* (NOS) estimate performance is strongly affected by the dynamics (*left*), but the estimate localizes the nondyadic jumps well (*right*); cf. Fig. 5.4

These convergence rates are similar to those in the nonlinear QOS algorithm, i.e., they are independent of:

- The type of system dynamics
- The correlation structure of the external noise (which can now have an unbounded density function)
- In contrast to the QOS algorithms,—independent of the smoothness of the probability density function of the input signal

5.6 Empirical Distribution Algorithm

In the *empirical distribution algorithm* (ED), we recover $\mu(u)$ estimating an equivalent *composite function* $\mu_F \circ F(u) = \mu_F(F(u))$, where $F(u)$ is the *cumulative distribution function* of the input signal (having a probability density function $f(u)$).

Let $\{(u_k, y_k)\}$, $k = 1, \ldots, N$ denote the set of the system input–output measurements sorted pairwise w.r.t. to the increasing values of the inputs u_k. The algorithm is based on two observations:

- The *empirical distribution,* $F_N(u)$, generated by such ordered inputs u_k, maps the randomly scattered sequence $\{u_k\}$ onto the equidistant grid $\{x_k\}$:

$$x_k = F_N(u_k) = \frac{k}{N}. \tag{5.32}$$

- By virtue of Assumption 1, the cumulative distribution function $F(u)$ is continuous, strictly increasing (viz., invertible).

Throughout the chapter, we assume, for simplicity, that N is a *dyadic integer* and hence $\{x_k\}$ forms a *dyadic grid* with grid points, $x_k = 2^{-\log_2 N}k$, being *binary rationals*.

Fig. 5.8 The illustration of
the idea behind the ED
algorithm. Note that input
measurements are equidistant
and each empirical coefficient
is computed using the same
number of measurements
(weighted by the same factor
$1/N$)

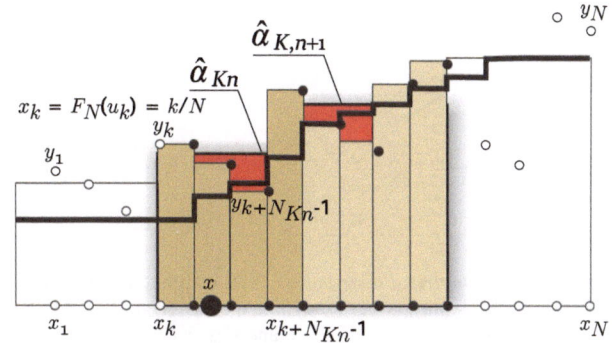

The *empirical distribution estimate* of the nonlinearity $\mu(x)$ is of the form:

$$\hat{\mu}_K(x) = \sum_{n=0}^{2^K-1} \hat{\alpha}_{Kn} \varphi_{Kn}(x) \quad \text{where } x = F_N(u), \tag{5.33}$$

and where the empirical coefficients $\hat{\alpha}_{Kn}$ are the following estimates

$$\hat{\alpha}_{Kn} = \sum_{k=1}^{N} y_k \int_{\frac{k-1}{N}}^{\frac{k}{N}} \varphi_{Kn}(x)\,\mathrm{d}x \tag{5.34}$$

of the unknown coefficients α_{Kn} of the nonlinearity $\mu(x)$ approximation in the
space V_K; see Fig. 5.8 and cf. (4.10) and (5.24)):

$$\alpha_{Kn} = \int_0^1 \mu(x)\varphi_{Kn}(x)\,\mathrm{d}x = \sum_{k=1}^{N} \int_{\frac{k-1}{N}}^{\frac{k}{N}} \mu(x)\varphi_{Kn}(x)\,\mathrm{d}x. \tag{5.35}$$

The algorithm based on similar application of the empirical distribution function
has been proposed in the statistical literature; see [26]. Moreover, the empirical
coefficients used there are calculated using the formula

$$\check{\alpha}_{Kn} = \frac{1}{N} \sum_{k=1}^{N} y_k \varphi_{Kn}(x_k), \tag{5.36}$$

which seems to be simpler to compute than ours. Nevertheless, applying (like in OS
algorithm) the *fundamental theorem of calculus* (see, e.g., [1, Th. 5.3]), we have that

$$\int_{\frac{k-1}{N}}^{\frac{k}{N}} \varphi_{Kn}(x)\,\mathrm{d}x = \Phi_{Kn}\left(\tfrac{k}{N}\right) - \Phi_{Kn}\left(\tfrac{k-1}{N}\right),$$

where $\Phi_{Kn}(x)$ are *indefinite integrals* of the scaling functions, $\varphi_{Kn}(x)$, and the
resulting equivalent form of our estimate:

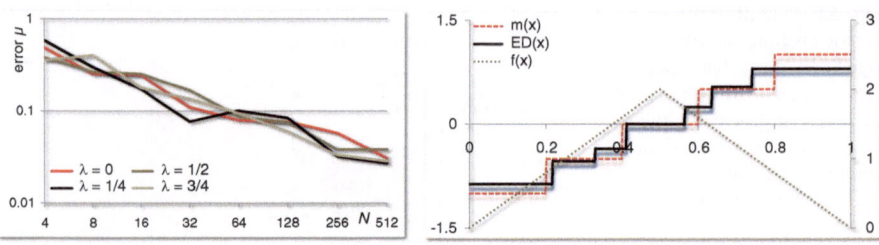

Fig. 5.9 The *empirical distribution* (ED) estimate is robust against the dynamics (*left*) but rather poorly localizes jumps at nondyadic points (*right*). Note the nonuniform, density-dependent estimate scale; cf. Fig. 5.2, Fig. 5.6 and Fig. 5.12

$$\hat{\alpha}_{Kn} = \sum_{k=1}^{N} y_k \left[\Phi_{Kn} \left(\tfrac{k}{N} \right) - \Phi_{Kn} \left(\tfrac{k-1}{N} \right) \right], \qquad (5.37)$$

is, in fact, equally simple since, in case of the Haar function, the indefinite integrals, $\Phi_{Kn}(x)$ have the common compact form $\Phi_{Kn}(x) = \sqrt{2^{-K}} \Phi \left(2^K x - n \right)$, where $\Phi(x) = x \cdot \chi_{[0,1)}(x) + \chi_{[1,\infty)}(x)$; cf. (5.25). Observing further that

$$\check{\alpha}_{Kn} = \sum_{k=1}^{N} y_k \int_{\frac{k-1}{N}}^{\frac{k}{N}} \varphi_{Kn}(x_k)\, dx,$$

we can expect in general a better performance of our estimate due to its more accurate approximation of the integral in (5.35); cf. Remark 5.6 in the previous chapter.[11]

Note that the number of measurements used to evaluate each of the ED algorithm empirical coefficient $\hat{\alpha}_{Kn}$ equals $2^{-K} N$ and thus is fixed, deterministic, and the same for all coefficients; cf. (5.34). In effect, the algorithm scale adapts to the local density of the input measurements (the local scale grows in regions where the number of measurement is larger than in the others; cf. Figs. 5.8, 5.9 and 7.1). This is a significant property which differentiates the ED algorithm from, e.g., the previous ones, where the number of measurements used to compute each of the empirical coefficients is random and changes from coefficient to coefficient; cf. (5.9) and (5.23).

5.6.1   Convergence

The mapping made by the empirical distribution $F_N(u)$ turns the randomly scattered inputs into the equidistant grid and may suggest that the initially random setting design regression estimation problem is turned into the fixed design one. Unfortunately, the presence of dynamics in the system (resulting in correlation

---

[11]For dyadic N, these formulas are actually equivalent; see Sect. 6.1.3.

of the output data) does not allow to directly apply the fixed design results (see, e.g., [67, 116, 121]) to establish the algorithm properties. Nonetheless, since the input measurements x_k are—in some sense—ordered versions of the raw inputs u_k, then the ED algorithm properties appear to be comparable to those possessed by the OS one.

The first theorem describes the convergence conditions of the algorithm.

Theorem 5.13. *If the scale K of the estimate in (5.33) is selected so that*

$$K \to \infty \text{ and } 2^K/N \to 0 \text{ as } N \to \infty,$$

then the empirical distribution algorithm converges to the nonlinearity $\mu(u)$ globally, in the MISE error sense, with growing number of measurements, i.e., we have

$$\text{MISE } \hat{\mu}_K \to 0 \text{ as } N \to \infty.$$

Likewise, the OS algorithm, the current one, converges globally to any piecewise-continuous nonlinearity for:

- Any piecewise-continuous input probability density function
- Any asymptotically stable system dynamics
- Any second-order noise (correlated or white)

5.6.2 Convergence Rates

In the theorems below, dealing with convergence rates, we assume the nonlinearity is Lipschitz or piecewise-Lipschitz. It is interesting to note that in spite of the fact that the nonlinearity is identified indirectly—via the composite function $\mu_F \circ F(u)$, the convergence rates are not dependent on the smoothness of the input probability density function.

Theorem 5.14. *Let the nonlinearity $\mu(u)$ be a Lipschitz function. If the scale K of the estimate in (5.33) is selected using the formula*

$$K = \tfrac{1}{3} \log_2 N, \tag{5.38}$$

then the empirical series algorithm converges to the nonlinearity globally, in the MISE error sense, with the rate

$$\text{MISE } \hat{\mu}_K = \mathcal{O}\left(N^{-2/3}\right).$$

The theorem implies that, similarly to the OS algorithm, the rate is independent of:

- The type of system dynamics
- The noise correlation structure
- The smoothness of the input probability density function

and depends only on smoothness of the nonlinearity $\mu(u)$. The next theorem is an analogue to Theorem 5.9 which demonstrated the slower convergence rate of the algorithm for $\mu(u)$ being discontinuous (piecewise-Lipschitz).

Theorem 5.15. *Let the nonlinearity $\mu(u)$ be a piecewise-Lipschitz function. If the scale K of the estimate in (5.33) is selected using the formula*

$$K = \tfrac{1}{2}\log_2 N, \tag{5.39}$$

then the empirical distribution algorithm converges to this nonlinearity, globally in the MISE error sense, with the rate

$$\text{MISE}\,\hat{\mu}_K = \mathcal{O}\left(N^{-1/2}\right).$$

The theorem reveals that also the ED algorithm convergences slower to the piecewise-Lipschitz nonlinearities than to the uniformly smooth ones. Observe further that even in the special case, when the nonlinearity $\mu(u)$ has jumps at dyadic points, the convergence rate of the ED algorithm does not accelerate to the $\mathcal{O}\left(N^{-2/3}\right)$—as it does for QOS and OS algorithms; cf. Theorems 5.4 and 5.10. This is because not the nonlinearity itself but the composite function $\mu_F(F(u))$, where the dyadic jump points are, in general, mapped by the distribution function into arbitrary locations, is identified. Clearly, the following corollary remains true for the ED algorithm; cf. Corollaries 5.1 and 5.2.

Corollary 5.3. *If the nonlinearity is piecewise-Lipschitz and the scale selection rule in (5.38) is used instead of that in (5.39), then the algorithm convergence rate is limited to $\mathcal{O}\left(N^{-1/3}\right)$.*

5.7   Nonlinear ED Algorithm

The nonlinear variant of the empirical distribution algorithm (NED) shares the EZW routine with the previous algorithms. It consists of the linear part with the empirical scaling functions coefficients $\hat{\alpha}_{Mn}$ and the nonlinear one, with the empirical wavelet coefficients $\hat{\beta}_{mn}$ selected with the help of EZW scheme. The resulting algorithm is therefore of the already familiar form (cf., e.g., (5.29))

$$\hat{\mu}_{MK}(x) = \sum_{n=0}^{2^M-1} \hat{\alpha}_{Mn}\varphi_{Mn}(x) + \sum_{m=M}^{K-1}\sum_{n\in Q_m} \hat{\beta}_{mn}\psi_{mn}(x), \tag{5.40}$$

with the empirical coefficients computed now as

$$\hat{\alpha}_{Mn} = \sum_{k=1}^{N} y_k \int_{\frac{k-1}{N}}^{\frac{k}{N}} \varphi_{Mn}(x)\,\mathrm{d}x \quad \text{and} \quad \hat{\beta}_{mn} = \sum_{k=1}^{N} y_k \int_{\frac{k-1}{N}}^{\frac{k}{N}} \psi_{mn}(x)\,\mathrm{d}x.$$

As in the previous algorithms, each subset Q_m stores q_M translation indices of that wavelet empirical coefficients $\hat{\beta}_{mn}$, which—according to the EZW algorithm—are situated in the cones of influences induced by the jumps of the identified nonlinearity.

5.7.1   Convergence and Convergence Rates

The first theorem describes the convergence conditions of the nonlinear ED algorithm.

Theorem 5.16. *If the scale M of the estimate in (5.40) and the empirical cone number q_M are selected so that*

$$\begin{cases} M \to \infty \\ q_M \to \infty \end{cases} and \quad \begin{array}{c} \frac{2^M}{N} \to 0 \\ \frac{q_M M}{2^M} \to 0 \end{array} , \quad as\ N \to \infty, \tag{5.41}$$

and the scale K is selected as
$$K = \nu M,$$

for any $\nu \geq 0$, then the nonlinear empirical distribution algorithm converges to the nonlinearity $\mu(x)$ globally, in the MISE error sense, with growing number of measurements, i.e., it holds that

$$\text{MISE}\ \hat{\mu}_{MK} \to 0\ as\ N \to \infty.$$

The theorem says that the convergence conditions of the nonlinear ED algorithm are actually the same as for the nonlinear QOS and OS versions, that is, the convergence properties are neither improved nor deteriorated by the presence of the EZW-generated nonlinear part. The restrictions in (5.41) imposed on q_M and K control the number and the "height" of the cones, respectively (cf. Fig. 4.2), and jointly ensure that the size of the nonlinear add-on (namely, the number of wavelet empirical coefficients) will not exceed the size of the linear basis.

In the next theorem the worst- and the best-case convergence rates of the ED algorithm are both established.

Theorem 5.17. *Let the nonlinearity $\mu(u)$ be piecewise-Lipschitz. If the scale M of the estimate in (5.40) is selected using the formula*

$$M = \tfrac{1}{3} \log_2 N,$$

while the scale K depends on M so that

$$K = 2M,$$

and the number of empirical cones is

$$q_M = \log_2 M, \tag{5.42}$$

then the nonlinear empirical distribution algorithm converges to this nonlinearity globally, in the MISE error sense, with the best-case rate

$$\text{MISE } \hat{\mu}_{MK} = \mathcal{O}\left(N^{-2/3}\right),$$

or with the worst-case rate

$$\text{MISE } \hat{\mu}_{MK} = \mathcal{O}\left(N^{-1/3}\right).$$

The convergence rates are thus similar to those in the previous QOS and OS nonlinear algorithms, i.e., they are robust against:

- The type of system dynamics
- The correlation structure of the external noise (which can now have an unbounded distribution)
- The smoothness of the probability density function of the input signal

5.8 Empirical Orthogonal Series Algorithm

The last identification algorithm considerably differs from all the previously presented: It is no longer based on the classic Haar basis but on the unbalanced Haar one (cf. Sect. 4.2).

Let $\{(x_k, y_k)\}$, $k = 1, \ldots, N$, be—like in the OS algorithm—the set of the measurement pairs obtained from the original measurements set $\{(u_k, y_k)\}$ by sorting the pairs (u_k, y_k) w.r.t. increasing input values (Fig. 5.10). Let further $\{\varphi_{Kn}(x)\}$, $n = 0, \ldots, 2^K - 1$, some $K < \log_2 N$, be the family of *the unbalanced Haar scaling function* generated *empirically* (adaptively) from the ordered measurement pairs; see formulas (4.43)–(4.44). Although the randomness of the input signal means that the supports (and the related normalization factors) of each basis function $\varphi_{Kn}(x)$ vary for different translation factors n and are unique for each realization of the measurements sets, *the EOS* algorithm is of rather familiar form, cf. (5.22) and (5.33)

$$\hat{\mu}_K(x) = \sum_{n=0}^{2^K-1} \hat{\alpha}_{Kn}\varphi_{Kn}(x) \text{ where } x = u, \tag{5.43}$$

where the empirical coefficients are given as follows (cf. (5.23) and (5.34)):

$$\hat{\alpha}_{Kn} = \sum_{k=1}^{N} y_k \int_{x_{k-1}}^{x_k} \varphi_{Kn}(x)\,dx. \tag{5.44}$$

Clearly, replacing the integration operations in (5.44) by subtractions of the indefinite integrals $\Phi_{Kn}(x)$ of the unbalanced Haar scaling functions $\varphi_{Kn}(x)$ leads to the equivalent integration-free formula (cf. (5.25) and (5.37))

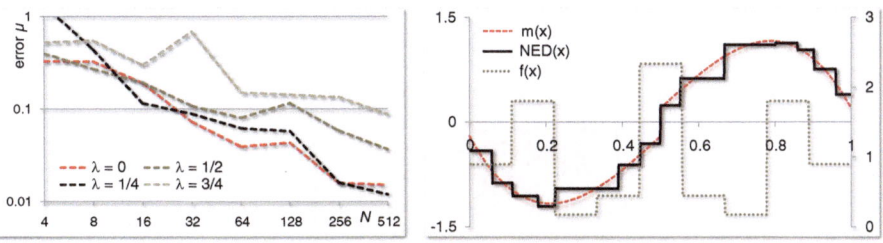

Fig. 5.10 The *nonlinear empirical distribution* (NED) estimate is heavily affected by the dynamics (*left*) but quite well represents the shape of the nolinearity (*right*); cf. Fig. 5.4, Fig. 5.7 and Fig. 5.13

$$\hat{\alpha}_{Kn} = \sum_{k=1}^{N} y_k \left[\Phi_{Kn}(x_k) - \Phi_{Kn}(x_{k-1}) \right].$$

Remark 5.7. Our estimate is a variant of the EOS one proposed in [58, Chap. 18.2], where the empirical coefficients are computed as follows (cf. also (5.36) in the ED algorithm):

$$\bar{\alpha}_{Kn} = \frac{1}{N} \sum_{k=1}^{N} y_k \varphi_{Kn}(x_k). \tag{5.45}$$

One can also consider another integration-free mutation of the formula in (5.44),

$$\check{\alpha}_{Kn} = \sum_{k=1}^{N} y_k (x_k - x_{k-1}) \varphi_{Kn}(x_k) = \sum_{k=1}^{N} y_k \int_{x_{k-1}}^{x_k} \varphi_{Kn}(x_k) \, dx,$$

which, however, may in general perform worse than the original one; cf. Remark 5.6.[12]

Note that the way the empirical coefficients $\hat{\alpha}_{Kn}$ in the EOS algorithm are computed resembles that in the OS one (where each output measurement y_k is weighted by the corresponding spacing $(x_k - x_{k-1})$); however, since the unbalanced orthogonal Haar basis is adapted to the input measurements, the number of output measurements taken into account by each coefficient routine is nonrandom (and equal to $2^{-K}N$ for all coefficients for the assumed dyadic N); see Fig. 5.11 and Sect. 6.1.4 and Appendix A.4. In consequence, the local scale of the algorithm is not fixed (as the form of the algorithm in (5.43) may suggest) but actually varies according to the local density of the input measurements; cf. Figs. 5.11, 5.12 and 7.1. In this aspect, the ED and EOS algorithms are much alike.

---

[12]For dyadic N, these formulas are actually equivalent; see Sects. 6.1.4 and 6.1.3.

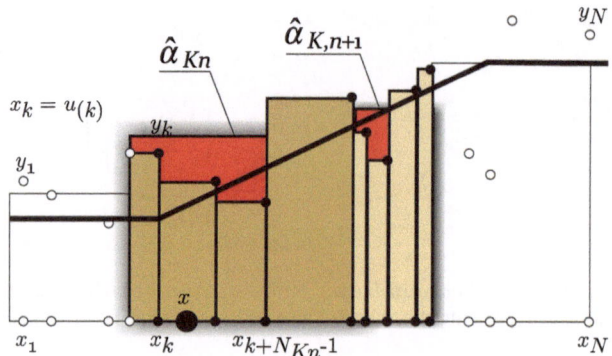

Fig. 5.11 The illustration of the idea behind the EOS algorithm. Note that input measurements are random, but each empirical coefficient is computed using the same number of measurements

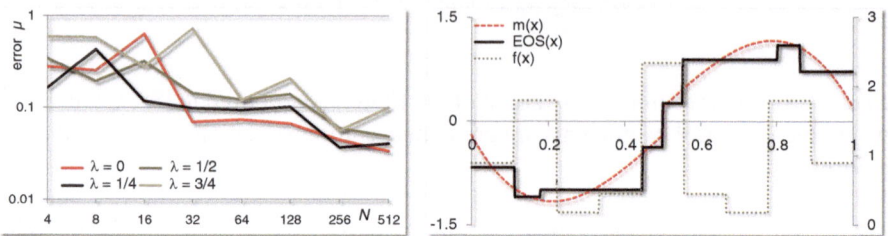

Fig. 5.12 The *empirical orthogonal series* (EOS) estimate is quite robust against the dynamics (*left*) but weakly restores the nonlinearity shape (*right*). Note the varying, density-dependent, estimate scale; cf. Fig. 5.9

5.8.1 Convergence

In spite of application of the unbalanced Haar family, the properties of the EOS algorithm are similar to those of previously presented algorithms. The only difference is that the convergence rates are slightly slower (by a logarithmic factor). The main reason for that is that the unbalanced orthogonal basis is generated by random inputs, and thus it is itself a collection of random (but still orthonormal) functions.

The following theorem describes the convergence conditions for the EOS algorithm.

Theorem 5.18. *If the scale K of the estimate in (5.43) is selected so that*

$$K \to \infty \text{ and } K^2 2^K / N \to 0 \text{ as } N \to \infty, \tag{5.46}$$

then the EOS algorithm converges to the nonlinearity $\mu(u)$ globally, in the MISE error sense, with growing number of measurements, i.e., we have

$$\text{MISE } \hat{\mu}_K \to 0 \text{ as } N \to \infty.$$

While the convergence condition in (5.46) are slightly different than for the previous algorithms (note the presence of the factor K^2), the EOS one converges globally to any piecewise-continuous nonlinearity for

- Any piecewise-continuous input probability density function
- Any asymptotically stable system dynamics
- Any second-order noise (correlated or white)

that is, its asymptotic properties can be expected to be similar to those of the OS and ED algorithms.

5.8.2 Convergence Rates

The following pair of theorems establishes the convergence rates of the algorithm for Lipschitz and piecewise-Lipschitz nonlinearities. Note that the rates are slightly slower (by a logarithmic factor) when compared to the rates attained by the previous algorithms.

Theorem 5.19. *Let the nonlinearity $\mu(u)$ be a Lipschitz function. If the scale K of the estimate in (5.43) is selected using the formula*

$$K = \tfrac{1}{3} \log_2 N,$$

then the empirical series algorithm converges to the nonlinearity globally, in the MISE error sense, with the rate

$$\mathrm{MISE}\,\hat{\mu}_K = \mathcal{O}\left(N^{-2/3} \log^3 N\right).$$

As for the OS and ED algorithms, the rate is independent of:

- The type of system dynamics
- The correlation structure of the external noise
- The smoothness of the input probability density function

and depends only on smoothness of the identified nonlinearity $\mu(u)$. The next theorem is a copy of the appropriate theorems for the OS and for the ED algorithms, for the case when $\mu(u)$ is discontinuous.

Theorem 5.20. *Let the nonlinearity $\mu(u)$ be piecewise-Lipschitz. If the scale K of the estimate in (5.43) is selected using the formula*

$$K = \tfrac{1}{2} \log_2 N$$

then the empirical distribution algorithm converges to this nonlinearity, globally in the MISE error sense, with the rate

$$\text{MISE } \hat{\mu}_K = \mathcal{O}\left(N^{-1/2} \log^3 N\right).$$

Recall that in the EOS algorithm, the basic functions $\varphi_{Kn}(x)$ have—by design—random supports. One consequence of this fact is that its convergence rate remains slower (like in the ED algorithm) even if the discontinuous nonlinearities have jumps at dyadic points (i.e., there are no counterparts of the "saturation" Corollaries 5.1 and 5.2 for the EOS algorithm).

5.9   Nonlinear EOS Algorithm

The nonlinear version of the empirical orthogonal series algorithm (NEOS), exploiting the common EZW nonlinear approximation scheme, also consists of the linear part with the empirical scaling functions coefficients, $\hat{\alpha}_{Mn}$, and of the nonlinear one, with the empirical wavelet coefficients selected adaptively by the EZW routine. The resulting algorithm is of the form

$$\hat{\mu}_{MK}(x) = \sum_{n=0}^{2^M-1} \hat{\alpha}_{Mn}\varphi_{Mn}(x) + \sum_{m=M}^{K-1} \sum_{n \in Q_m} \hat{\beta}_{mn}\psi_{mn}(x), \qquad (5.47)$$

with the Q_m being collection of the translation indices n corresponding, at each scale $m = M, \ldots, K-1$, with q_M coefficients inside the selected empirical influence cones. The empirical coefficients are computed as (cf. (5.44))

$$\hat{\alpha}_{Mn} = \sum_{k=1}^{N} y_k \int_{x_{k-1}}^{x_k} \varphi_{Mn}(x)\,dx \text{ and } \hat{\beta}_{mn} = \sum_{k=1}^{N} y_k \int_{x_{k-1}}^{x_k} \psi_{mn}(x)\,dx.$$

5.9.1   Convergence and Convergence Rates

The first theorem describes the convergence conditions of the nonlinear EOS algorithm (compare the analogous Theorem 5.16 derived for ED algorithm).

Theorem 5.21. *If the scale M of the estimate in (5.47) and the cone number q_M are selected so that*

$$\begin{cases} M \to \infty \\ q_M \to \infty \end{cases} \text{ and } \begin{array}{l} \frac{2^M}{N}M^2 \to 0 \\ \frac{q_M}{2^M}M \to 0 \end{array} \text{ , as } N \to \infty, \qquad (5.48)$$

and the scale K is selected as

$$K = \nu M,$$

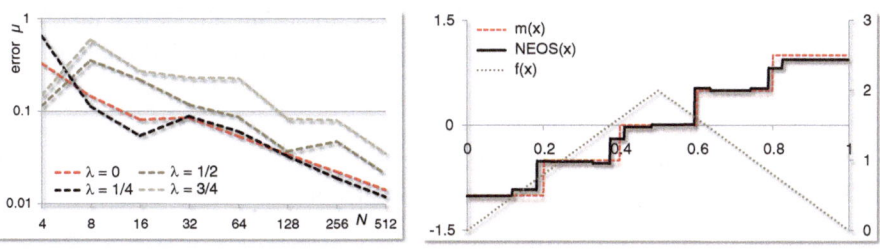

Fig. 5.13 The *nonlinear empirical orthogonal series* (NEOS) estimate is strongly affected by the dynamics (*left*) but very localizes nondyadic jumps (*right*); cf. Fig. 5.10

for any $v \geq 0$, then the nonlinear empirical orthogonal series algorithm converges to the nonlinearity $\mu(x)$ globally, in the MISE error sense, with growing number of measurements, i.e., it holds that

$$\text{MISE } \hat{\mu}_{MK} \to 0 \text{ as } N \to \infty.$$

The convergence conditions established for the nonlinear EOS algorithm are the same as for the nonlinear ED variant, and all the respective comments concerning the selection of the scales M, K, and cone number q_M are valid. The last theorem describes the worst- and the best-case convergence rates of the algorithm.

Theorem 5.22. *Let the nonlinearity $\mu(u)$ be piecewise-Lipschitz. If the scale M of the estimate in (5.47) is selected using the formula*

$$M = \tfrac{1}{3} \log_2 N,$$

and

$$K = 2M,$$

and the number of empirical cones is

$$q_M = \log_2 M, \tag{5.49}$$

then the nonlinear empirical orthogonal series algorithm converges to this nonlinearity globally, in the MISE error sense, with the best-case rate

$$\text{MISE } \hat{\mu}_{MK} = \mathcal{O}\left(N^{-2/3} \log^3 N\right),$$

or with the worst-case rate

$$\text{MISE } \hat{\mu}_{MK} = \mathcal{O}\left(N^{-1/2} \log^3 N\right).$$

The convergence rates of the nonlinear EOS algorithm are slower than those obtained for the rest of the algorithms; however, they remain (cf. Assumptions 1–4) not influenced by:

- The type of system dynamics
- The correlation structure of the external noise
- The smoothness of the probability density function of the input signal

5.10   Remarks and Comments

The first two linear identification algorithms, QOS and OS, are the Haar wavelet versions of the existing algorithms exploiting nonparametric regression estimates and based on either kernel or on classic (trigonometric or polynomial) orthogonal series functions; see [57, 68, 69, 110]. The other two, ED and EOS, are new and inspired by some techniques proposed in the statistical literature; see [26, 58]. The nonlinear variants of all algorithms are all new.

5.10.1   Convergence Rates Comparison

In Table 5.1, the (best-case) convergence rates of the proposed algorithms are summarized. In the worst-case scenarios, all rates slow down to $\mathcal{O}\left(N^{-1/3}\right)$ (or to $\mathcal{O}(N^{-1/3}\log^3 N)$ for NEOS algorithm) when the nonlinearity is piecewise-Lipschitz.

5.10.2   Common Representation

At the beginning of the chapter, we have stated that our algorithms are derived from the common *local averaging paradigm*, and—in the subsequent chapters—we have implemented that paradigm in the proposed algorithms constructions. Somehow naturally, the resulting estimates can therefore be represented by a single multiscale formula, viz., as the *regressograms* of the identified nonlinearity; cf. Sect. 4.1.5. This is particularly easy to observe in case of the linear algorithms which—for the measurement pairs $\{(x_k, y_k)\}$ ordered w.r.t. ascending input values—share the same generic form (see (5.6), (5.22), (5.33), (5.43) and Figs. 5.1, 5.5, 5.8 and 5.11, and cf. [58, 68, 69, Chap. 2.1])[13]

---

[13]Recall, that in the EOS case, $\varphi_{Kn}(x)$ stands for the scaling functions of the unbalanced Haar basis. Also in all algorithms $x = u$ but $x = \hat{F}_N(u)$ in the ED one.

Table 5.1 The (best-case) convergence rates of the algorithms

	Lipschitz		
	Uniform	Piecewise	
		Dyadic	Non-dyadic
QOS OS	$\mathcal{O}\left(N^{-2/3}\right)$		$\mathcal{O}\left(N^{-1/2}\right)$
ED	$\mathcal{O}\left(N^{-2/3}\right)$	$\mathcal{O}\left(N^{-1/2}\right)$	
EOS	$\mathcal{O}\left(N^{-2/3}\log^2 N\right)$	$\mathcal{O}\left(N^{-1/2}\log^2 N\right)$	
NQOS NOS NED	$\mathcal{O}\left(N^{-2/3}\right)$		
NEOS	$\mathcal{O}\left(N^{-2/3}\log^2 N\right)$		

$$\hat{\mu}_K(x) = \sum_{n=0}^{2^K-1} \hat{\alpha}_{Kn}\varphi_{Kn}(x) \quad \text{with} \quad \hat{\alpha}_{Kn} = \sum_{k=1}^{N} y_k \cdot \phi_{Kn,k}, \qquad (5.50)$$

but have only different weighting factors $\phi_{Kn,k}$ in their empirical coefficients formulas[14]:

$$\phi_{Kn,k} = \begin{cases} \int_{\frac{k-1}{N_{Kn}}}^{\frac{k}{N_{Kn}}} \varphi_{Kn}(x_k)\,dx, & x_k = u_k \quad (5.7)\text{---QOS,} \\[2ex] \int_{x_{k-1}}^{x_k} \varphi_{Kn}(x)\,dx, & x_k = u_{(k)} \quad (5.23)\text{---OS,} \\[2ex] \int_{x_{k-1}}^{x_k} \varphi_{Kn}(x)\,dx, & x_k = \frac{k}{N} \quad (5.34)\text{---ED,} \\[2ex] \int_{x_{k-1}}^{x_k} \varphi_{Kn}(x)\,dx, & x_k = u_{(k)} \quad (5.44)\text{---EOS.} \end{cases} \qquad (5.51)$$

Analogously, the nonlinear algorithms can have a common multiscale representation similar to that in (5.50)–(5.51). Nevertheless, in the nonlinear algorithms, the scale factor is no longer fixed but varies between the M and K and is locally determined by the EZW scheme (viz., it is dependent on the measurement data at hand; cf. (4.61) and (A.10)–(A.11))[15]

---

[14] Where $u_{(k)}$, $k = 1,, N$, is the kth input measurement after ordering (sorting) the raw measurements set $\{(u_k, y_k)\}$.

[15] The existence of this representation is essential for the properties of the nonlinear version of the QOS algorithm and helpful in the EOS algorithm analysis; see Appendices A.1 and A.4.

$$\hat{\mu}_{MK}(x) = \sum_{n=0}^{2^K-1} \hat{\alpha}_{K(n),n} \varphi_{K(n),n}(x), \ K(n) \in \{M, \dots, K\} \qquad (5.52)$$

with accordingly modified empirical coefficients formulas.

Remark 5.8. One can further recognize the similarities between the nonlinear algorithm's representation in (5.52) and *radial basis function*-based *(RBF)* estimates (cf., e.g., [37, 84, 91]; see also, e.g., [11, 12], for the surveys, and [9], where the relations between wavelets and RBFs are thoroughly examined).

5.10.3   Application of Interpolation Schemes

The generic multiscale representation in (5.50) can also be interpreted as the first-order (or zero-degree B-spline) interpolation of the identified nonlinearity where the interpolation coordinates $\{(x_n, \hat{\mu}_K(x_n))\}, n = 0, \dots, 2^K - 1$, are generated by the interpolation knots, $x_n = 2^{-K}(n + 1/2)$, located in the middle of the scaling functions $\varphi_{Kn}(x)$ supports and the filtered (averaged by any of the identification algorithm $\hat{\mu}_K(x)$) output observations; cf. [28, 112]. One can thus easily ascertain (cf., e.g., [70]) that these coordinates may serve as a basis for a higher-order interpolation scheme (see, e.g., [8, 100, 144])—if the direct estimate output (i.e., a discontinuous function) is not appropriate.

Example 5.1. A piecewise-linear (second-order) interpolation offers a simple (yet continuous) representation of the target nonlinearity and simultaneously maintains our algorithms' localization properties without a significant computational overhead. Such an interpolation scheme—written in the vein of (5.50)—has the following formula

$$\hat{\mu}_{K,\text{int}}(x) = \sum_{n=0}^{2^K-1} \hat{\mu}_K(x_n) \Lambda_{Kn}(x) \ \text{with} \ \hat{\mu}_K(x_n) = 2^{\frac{K}{2}} \hat{\alpha}_{Kn}, \qquad (5.53)$$

where $\Lambda_{Kn}(x) = \Lambda(2^K x - n)$ are contracted and translated (respectively, by factors 2^K and n) versions of the well-known linear B-spline *(hat/tent function)*, $\Lambda(x) = \chi_{[0,1]}(1 - |x|)$. The Fig. 5.14 compares the basis ED estimate $\hat{\mu}_K(x)$ and the built upon it interpolant $\hat{\mu}_{K,\text{int}}(x)$ for the piecewise-polynomial nonlinearity (5.2). Note that the knots do not form an equidistant grid.

Higher-order interpolation schemes—popular in image processing applications (like, e.g., cubic splines, [86, 144])—should carefully be used as they could introduce the *Gibbs-like oscillation* artifacts if the identified nonlinearities are discontinuous (cf., e.g., [144, 146, 147, Chap. 8]). Also, the classic *Lagrange* interpolation, due to its nonlocal nature and poor performance at boundaries (caused

Fig. 5.14 Combination of the Haar estimate (here of the ED type) and linear interpolation *(lerp)* can be beneficial when the identified nonlinearity is smooth (e.g., polynomial of higher order) between jumps

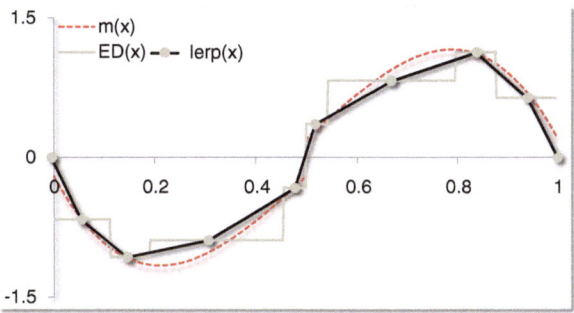

by the *Runge's oscillations*, see e.g., [21, Chap. 4.3.4] and [98, Chap. 6.1]), should rather not be allowed for; cf. also [113, Chap. 3.0].

Remark 5.9. The convergence rate of the estimation-interpolation assembly in (5.53) is—for all smooth interpolation schemes—limited by the approximation capabilities of the Haar basis, and application of higher-order interpolations does not improve the asymptotic behavior (convergence rate) of such composite estimates (cf., e.g., [145]).

Chapter 6
Computational Algorithms

Abstract Effective (that is, fast and involving only elementary arithmetic and comparison operations) computational counterparts of the identification algorithms are proposed. The implementations of the linear and nonlinear variants are typically split into the analysis and synthesis phases, and both are substantially based on the lifting wavelet transforms. The complexity of the algorithms is examined and compared.

It is widely recognized that "good identification algorithms" should possess the following prerequisites:

- Working for a broad class of characteristics of the system components, the input signals and noises
- Offering the effective (parsimonious) representations of these characteristics
- Having a form which can easily be evaluated from random measurements.

Clearly, these prerequisites are not exclusive for system identification algorithms. The similar postulates have been formulated in the renowned paper [36], by Donoho et al., in the related context of data compression:

> There is a **"Grand Challenge"** facing the disciplines of both theoretical and practical data compression in the future: the challenge of dealing with the particularity of naturally occurring phenomena. This challenge has three facets:
>
> - Obtaining accurate models of naturally occurring sources of data
> - Obtaining "optimal representations" of such models
> - Rapidly computing such "optimal representations"

In the previous chapter, we have proposed algorithms which, by construction, comply with the first two postulates: The nonparametric assumptions meet the first requirement, while the application of wavelets allows successful implementation of the second one. Nevertheless, the specific system identification assumptions (in particular, the randomness of the input signal) complicate the adoption of the fast wavelet transforms, i.e., the fast wavelet coefficient computation algorithms commonly used in signal and image processing tasks (see e.g., [24, 96] and cf. [41, 131, 132]). Hence, to make our identification algorithms comply with the third requirement, in the subsequent chapters, we develop and examine such 'rapid' computational implementations.

6.1   Linear and Nonlinear Algorithms Implementations

The essential difference between the implementations of the linear and nonlinear algorithms is that in the former the empirical scaling function coefficients are only computed, while in the latter both the empirical scaling function and wavelet coefficients need to be evaluated. Nevertheless, all—linear and nonlinear—algorithm implementations share a common two-phase template (they are often referred to as the *analysis* and *synthesis* phases in the wavelet literature—see, e.g., [24, 27, 118]): The first phase consists in computing the empirical coefficients, while in the second, the estimates outputs are evaluated. Eventually, all these routines produce tabulated values of the estimate outputs, i.e., the regressograms of the identified nonlinearities.

The implementation design takes into account both the order of their computational complexity, and the number of the algorithms passes over the data. In particular, to reduce the number of passes, the nonlinear approximation EZW scheme is 'sandwiched' between the linear algorithm analysis and synthesis phases and implemented as a *plug-in* procedure intertwined with the forward and inverse transforms steps.

6.1.1   QOS Algorithm

In the following section we propose a fast *one-pass* implementation of the QOS algorithm. The implementation does not require the measurement data to be ordered.

Analysis Phase

Let $\hat{\alpha}_{Kn}^{(k)}$ and $\hat{a}_{Kn}^{(k)}$ denote the empirical coefficients computed for k measurements. The proposed routine consists of two steps performed for each measurement pair (x_k, y_k), $k = 1, \ldots, N$:

- The translation index, n, of the coefficients $\hat{\alpha}_{Kn}^{(k)}$ and $\hat{a}_{Kn}^{(k)}$ affected by the current pair (x_k, y_k), is calculated as

$$n = \lfloor 2^K x_k \rfloor,$$

- The selected coefficients $\hat{\alpha}_{Kn}^{(k)}$ and $\hat{a}_{Kn}^{(k)}$ are updated by the trivial recurrence formula

$$
\begin{bmatrix} \hat{\alpha}_{Kn}^{(k)} \\ \hat{a}_{Kn}^{(k)} \end{bmatrix} = \begin{bmatrix} \hat{\alpha}_{Kn}^{(k-1)} \\ \hat{a}_{Kn}^{(k-1)} \end{bmatrix} + \begin{bmatrix} y_k \\ 1 \end{bmatrix},
\tag{6.1}
$$

where, again, the matrix-like form is used for compactness of the notation and to emphasize the similarity between the estimate numerator and denominator.

The formulas in (6.1) are simply the recursive versions of the summation formulas in (5.7); cf. (5.9). Note also that the weighting factor N (and the implicit scaling factor $2^{K/2}$), present in (5.7), can be omitted due to the quotient form of the estimate.

Complexity

Both algorithm steps are performed in a constant time and hence have the cost $\mathcal{O}(1)$. The overall algorithm complexity is thus $\mathcal{O}(N)$, i.e., it is linear w.r.t. the number of measurements N.

Remark 6.1. While the measurements need not to be sorted, the first step of the algorithm—determining the coefficients to be updated—can be seen as a version of bucket sorting (without the in-bucket ordering step) (cf., e.g., [89, Chap. 5.2.5]).

Synthesis Phase

The following formula (cf. (5.6) and (5.9))

$$
\hat{\mu}_K(x) = \left. \frac{\hat{\alpha}_{Kn}}{\hat{a}_{Kn}} \right|_{n = \lfloor 2^K x \rfloor},
$$

can be used in fast (i.e., with the constant cost $\mathcal{O}(1)$) calculations of the QOS estimate output for a given argument x.

6.1.2 OS Algorithm

Introducing the OS algorithm in Sect. 5.4, we have already suggested a possible simplification of the empirical coefficients computation routine (i.e., the analysis phase) by replacing integrations with subtractions; cf. (5.23) and (5.25). In order to further reduce the computation burden, we propose here a convenient recursive *one-pass* implementation of that phase (where, in particular, the raw measurement data set $\{(u_k, y_k)\}$, $k = 1, \ldots, N$, is not ordered in a separate routine).

Analysis Phase

Initially, the set of algorithm data measurements $\{(x_k, y_k)\}$ consists of two artificial 'boundary' measurement pairs (cf. Sect. 5.4)

$$\{(x_{-1}, y_{-1}), (x_0, y_0)\} = \{(0,0), (1,0)\},$$

and all empirical coefficients are zero, i.e., $\hat{\alpha}_{Kn}^{(0)} = 0$. Denote now by $\hat{\alpha}_{Kn}^{(k)}$ the empirical coefficient $\hat{\alpha}_{Kn}$ obtained for $k = 1, \ldots, N$ measurements.

Given the sequence ordered w.r.t. inputs, $\{(x_1, y_1), \ldots, (x_l, y_l), (x_{l+1}, y_{l+1}), \ldots, (x_k, y_k)\}$, assume that the $(k + 1)$th measurement pair (u_{k+1}, y_{k+1}) (from the raw measurement set) falls between the measurement pairs (x_l, y_l) and (x_{l+1}, y_{l+1}), i.e., assume that $x_l < u_{k+1} < x_{l+1}$. Then,

- The new data pair $(x_{k+1} = u_{k+1}, y_{k+1})$ is created and inserted between the pairs (x_l, y_l) and (x_{l+1}, y_{l+1}) to maintain the ascending order of the updated measurement set.
- The translation index n of the empirical coefficient $\hat{\alpha}_{Kn}^{(k)}$ to be updated is selected as

$$n = \lfloor 2^K x_{k+1} \rfloor.$$

Eventually,
- The following recurrence formula is applied to this coefficient:

$$\hat{\alpha}_{Kn}^{(k+1)} = \hat{\alpha}_{Kn}^{(k)} + (y_{k+1} - y_{l+1}) \left[\Phi_{Kn}(x_{k+1}) - \Phi_{Kn}(x_l) \right]. \quad (6.2)$$

The formula in (6.2) can easily be derived by subtraction of the estimate in (5.25), computed for k measurements, from the one obtained for $k + 1$ measurements and by the subsequent rearrangement of the residue terms (cf., e.g., [133]). Indeed, from

$$\hat{\alpha}_{Kn}^{(k)} = \sum_{i=1}^{l} y_i \left[\Phi_{Kn}(x_i) - \Phi_{Kn}(x_{i-1}) \right] + \sum_{i=l+2}^{k} y_i \left[\Phi_{Kn}(x_i) - \Phi_{Kn}(x_{i-1}) \right]$$
$$+ y_{l+1} \left[\Phi_{Kn}(x_{l+1}) - \Phi_{Kn}(x_l) \right]$$

and

$$\hat{\alpha}_{Kn}^{(k+1)} = \sum_{i=1}^{l} y_i \left[\Phi_{Kn}(x_i) - \Phi_{Kn}(x_{i-1}) \right] + \sum_{i=l+2}^{k} y_i \left[\Phi_{Kn}(x_i) - \Phi_{Kn}(x_{i-1}) \right]$$
$$+ y_{k+1} \left[\Phi_{Kn}(x_{k+1}) - \Phi_{Kn}(x_l) \right] + y_{l+1} \left[\Phi_{Kn}(x_{l+1}) - \Phi_{Kn}(x_{k+1}) \right],$$

we get

$$\hat{\alpha}_{Kn}^{(k+1)} - \hat{\alpha}_{Kn}^{(k)} = y_{k+1} \left[\Phi_{Kn}(x_{k+1}) - \Phi_{Kn}(x_l) \right]$$
$$+ y_{l+1} \left[\Phi_{Kn}(x_{l+1}) - \Phi_{Kn}(x_{k+1}) \right]$$
$$- y_{l+1} \left[\Phi_{Kn}(x_{l+1}) - \Phi_{Kn}(x_l) \right],$$

which, after terms regrouping, yields the required recurrence relation

$$\hat{\alpha}_{Kn}^{(k+1)} - \hat{\alpha}_{Kn}^{(k)} = y_{k+1}\left[\Phi_{Kn}\left(x_{k+1}\right) - \Phi_{Kn}\left(x_l\right)\right]$$
$$+ y_{l+1}\left[\Phi_{Kn}\left(x_l\right) - \Phi_{Kn}\left(x_{k+1}\right)\right]$$
$$= \left(y_{k+1} - y_{l+1}\right)\left[\Phi_{Kn}\left(x_{k+1}\right) - \Phi_{Kn}\left(x_l\right)\right].$$

Complexity

The insertion step has the complexity $\mathcal{O}\left(\log k\right)$ when the self-balancing *AVL* or *red-black trees* are used to store the measurement set $\{(x_k, y_k)\}$; cf. [89] and [3]. Also the formula in (6.2) requires $\mathcal{O}\left(\log k\right)$ operations to find the adjacent pairs (x_l, y_l) and (x_{l+1}, y_{l+1}) in such trees. The complexity of the analysis phase of the OS algorithm is thus $\mathcal{O}\left(N \log N\right)$, i.e., it is larger by a log factor than the complexity of this phase in the QOS algorithm.

Synthesis Phase

Given the set of empirical coefficients $\hat{\alpha}_{Kn}$, the OS estimate reduces, for each argument x, to the form

$$\hat{\mu}\left(x\right) = 2^{\frac{K}{2}}\hat{\alpha}_{Kn}\Big|_{n=\lfloor 2^K x \rfloor},$$

i.e., to find the output of the estimate for a given x, it suffices to take the value of the empirical coefficient $\hat{\alpha}_{Kn}$ (with the translation index $n = \lfloor 2^K x \rfloor$) and multiply it by the scaling factor $2^{K/2}$. The OS algorithm synthesis phase complexity is thus $\mathcal{O}\left(1\right)$ and equals the complexity of this phase in the QOS algorithm.

6.1.3   ED Algorithm

In order to present the fast implementation of the ED algorithm—which will be based on direct application of the standard fast wavelet transform (FWT) in (4.26)—we need some preliminary derivations.

First, we need to show that, given the empirical coefficients at the scale m, the fast wavelet transform can be used to compute the empirical coefficients at the scale $m - 1$ (cf. (4.4) and (4.26)):

$$\hat{\alpha}_{m-1,n} = \sum_{k=1}^{N} y_k \int_{\frac{k-1}{N}}^{\frac{k}{N}} \varphi_{m-1,n}\left(x\right) dx$$

$$= \frac{1}{\sqrt{2}} \sum_{k=1}^{N} y_k \int_{\frac{k-1}{N}}^{\frac{k}{N}} \left[\varphi_{m,2n}\left(x\right) + \varphi_{m,2n+1}\left(x\right)\right] dx$$

$$= \frac{1}{\sqrt{2}} \sum_{k=1}^{N} y_k \int_{\frac{k-1}{N}}^{\frac{k}{N}} \varphi_{m,2n}(x)\, dx + \frac{1}{\sqrt{2}} \sum_{k=1}^{N} y_k \int_{\frac{k-1}{N}}^{\frac{k}{N}} \varphi_{m,2n+1}(x)\, dx$$

$$= \frac{1}{\sqrt{2}} \left(\hat{\alpha}_{m,2n} + \hat{\alpha}_{m,2n+1} \right). \tag{6.3}$$

and that, for dyadic N, the properly scaled versions of the output measurements y_k can serve as the initial empirical coefficients $\hat{\alpha}_{\log_2, n}$ for the fast wavelet transform, i.e., that

$$\hat{\alpha}_{\log_2 N, n} = 2^{\frac{\log_2 N}{2}} y_{n+1} = \frac{\sqrt{N}}{N} y_{n+1}. \tag{6.4}$$

Exploiting now the piecewise-constant form of Haar scaling functions and the compactness of their supports (4.7), we infer that for each n and each $m \le \log_2 N$, the formula (5.34) is equivalent to the (scaled) sums of $2^{-m} N$ consecutive output measurements y_k, $k = 2^{-m} N n + 1, \ldots, 2^{-m} N (n+1)$[1]:

$$\hat{\alpha}_{mn} = \sum_{k=1}^{N} y_k \int_{\frac{k-1}{N}}^{\frac{k}{N}} \varphi_{mn}(x)\, dx = \sum_{k=2^{-m} N n + 1}^{2^{-m} N(n+1)} y_k \int_{\frac{k-1}{N}}^{\frac{k}{N}} \varphi_{mn}(x)\, dx$$

$$= \sum_{k=2^{-m} N n + 1}^{2^{-m} N(n+1)} y_k \varphi_{mn}\left(\tfrac{k-1}{N}\right) \int_{\frac{k-1}{N}}^{\frac{k}{N}} dx = \frac{1}{N} \sum_{k=2^{-m} N n + 1}^{2^{-m} N(n+1)} y_k \varphi_{mn}\left(\tfrac{k-1}{N}\right)$$

$$= \frac{2^{\frac{m}{2}}}{N} \sum_{k=2^{-m} N n + 1}^{2^{-m} N(n+1)} y_k \tag{6.5}$$

To verify (6.4), we examine the empirical coefficients $\hat{\alpha}_{mn}$ at the scale $m = \log_2 N$. Combining (6.3) and (6.5), we get for this scale (since $2^{-\log_2 N} N = 2$) that:

$$\hat{\alpha}_{\log_2 N-1, n} = \frac{1}{\sqrt{2}} \frac{\sqrt{N}}{N} \sum_{k=2n+1}^{2(n+1)+1} y_k = \frac{1}{\sqrt{2}} \frac{\sqrt{N}}{N} (y_k + y_{k+1})$$

$$= \frac{1}{\sqrt{2}} \left(\frac{\sqrt{N}}{N} y_{2n+1} + \frac{\sqrt{N}}{N} y_{2n+2} \right)$$

$$= \frac{1}{\sqrt{2}} \left(\hat{\alpha}_{\log_2 N, 2n} + \hat{\alpha}_{\log_2 N, 2n+1} \right). \tag{6.6}$$

Analysis Phase

The fast implementation of the analysis phase of the ED algorithm is now rather straightforward and consists of the following three steps:

---

[1] Observe that, for dyadic N, also the formulas (5.34) and (5.36) are equivalent.

- Sorting the set of the raw measurement pairs $\{(u_k, y_k)\}$ w.r.t. the ascending values of the input measurements in order to construct the empirical distribution $F_N(u)$ of the input measurements.
- Performing the fast wavelet transform routine on the output measurements scaled by a factor $\sqrt{N^{-1}}$. The routine starts from the scale $m = \log_2 N - 1$ and is repeated down to $m = K$.
- Storing every $2^{-K}N$th entry of the original empirical distribution $F_N(u)$ as a new map function $F_K(u)$.

The function $F_K(u)$ constructed in the last step consists of 2^K entries and is used in the subsequent synthesis phase to map the input arguments u onto the translation indexes n of the corresponding to these arguments empirical coefficients $\hat{\alpha}_{Kn}$.

Complexity

The first step consists in sorting of the measurements w.r.t. the ascending input values and requires $\mathcal{O}(N \log N)$ operations when, e.g., *heapsorting* algorithm is employed (cf. e.g., [89, Chap. 5.2.3]). To verify that the second step needs only $\mathcal{O}(N)$ operations, observe that the number of operations at each scale m is of order $\mathcal{O}(2^m)$ since the single operation in (4.26) is performed for $n = 0, \ldots, 2^m - 1$ times. To find the empirical coefficients at the scale $m = K \geq 0$, we thus need to perform

$$\sum_{m=K}^{\log_2 N - 1} \mathcal{O}(2^m) = \mathcal{O}\left(2^{\log_2 N} - 2^K\right) = \mathcal{O}(N)$$

operations; cf. [96, Chap. 7.3.1]. Construction of the map function $F_K(x)$ requires $\mathcal{O}(2^K) = \mathcal{O}\left(\sqrt[3]{N}\right)$ operations. This yields $\mathcal{O}(N \log N)$ as the overall cost of the analysis phase.

Synthesis Phase

The synthesis phase implements the following formula:

$$\hat{\mu}(u) = 2^{\frac{K}{2}} \hat{\alpha}_{Kn}\Big|_{n = \lfloor 2^K F_K(u) \rfloor},$$

and the estimate output can be thus computed in two simple steps:

- Determining the translation index n for a given argument u using the map $F_K(u)$
- Evaluating the output value, i.e., scaling the selected coefficient $\hat{\alpha}_{Kn}$ by the factor $2^{K/2}$

Remark 6.2. The second step can be performed as the last one in the analysis phase as well.

Complexity

The first step can be implemented using a *binary search algorithm*, which for the sets of 2^K elements requires $\mathcal{O}(K)$ operations; cf., e.g., [89, Chap. 6.2.1]. Since $K = 1/3 \log_2 N$ and the evaluation of the output is clearly an $\mathcal{O}(1)$ operation, we conclude that the overall cost of the synthesis phase of the ED algorithm, being of order

$$\mathcal{O}\left(\log \sqrt[3]{N}\right) = \mathcal{O}(\log N),$$

is larger than the constant one obtained for QOS and OS implementations.

6.1.4 EOS Algorithm

The implementation of the EOS algorithm is similar to the ED's one, and we start with showing that the analysis phase (i.e., computing the empirical coefficients $\hat{\alpha}_{Kn}$) can be performed by a direct application of the unbalanced Haar transform (UHT) to the properly scaled measurement data from the pairwise-ordered set $\{(x_k, y_k)\}$.

In the following preparatory step, we will demonstrate that the unbalanced transform can be used to compute empirical coefficients at the scale $m - 1$ given the coefficients at the scale m (cf. (4.58) and (6.3)):

$$\hat{\alpha}_{m-1,n} = \sum_{k=1}^{N} y_k \int_{x_{k-1}}^{x_k} \varphi_{m-1,n}(x)\,dx$$

$$= \sum_{k=1}^{N} y_k \int_{x_{k-1}}^{x_k} \left[\frac{\sqrt{I_{m,2n}}}{\sqrt{I_{m-1,n}}} \varphi(x)_{m,2n} + \frac{\sqrt{I_{m,2n+1}}}{\sqrt{I_{m-1,n}}} \varphi(x)_{m,2n+1} \right] dx$$

$$= \frac{\sqrt{I_{m,2n}}}{\sqrt{I_{m-1,n}}} \sum_{k=1}^{N} y_k \int_{x_{k-1}}^{x_k} \varphi(x)_{m,2n}\,dx$$

$$+ \frac{\sqrt{I_{m,2n+1}}}{\sqrt{I_{m-1,n}}} \sum_{k=1}^{N} y_k \int_{x_{k-1}}^{x_k} \varphi(x)_{m,2n+1}\,dx$$

$$= \frac{\sqrt{I_{m,2n}}}{\sqrt{I_{m-1,n}}} \hat{\alpha}_{m,2n} + \frac{\sqrt{I_{m,2n+1}}}{\sqrt{I_{m-1,n}}} \hat{\alpha}_{m,2n+1}, \tag{6.7}$$

where I_{mn} stands for the length of the support of the unbalanced scaling function $\varphi_{mn}(x)$ as in (4.44). To prove that the scaled measurement data can be used as the initial empirical coefficients for the unbalanced transform, i.e., that

$$\hat{\alpha}_{\log_2 N, n} = \sqrt{I_{\log_2 N, n}}\, y_{n+1} = \sqrt{I_{n+1}}\, y_{n+1}, \tag{6.8}$$

we need first to show that the following equivalence holds for dyadic N and for each $m \leq \log_2$ and each n:

$$\hat{\alpha}_{mn} = \sum_{k=2^{-m}Nn+1}^{2^{-m}N(n+1)} y_k \int_{x_{k-1}}^{x_k} \varphi_{mn}(x)\,dx = \frac{1}{\sqrt{I_{mn}}} \sum_{k=2^{-m}Nn+1}^{2^{-m}N(n+1)} y_k \int_{x_{k-1}}^{x_k} dx$$

$$= \frac{1}{\sqrt{I_{mn}}} \sum_{k=2^{-m}Nn+1}^{2^{-m}N(n+1)} I_k y_k, \tag{6.9}$$

that is, that the empirical unbalanced scaling function coefficients $\hat{\alpha}_{mn}$ are scaled (by the normalization factor $1/\sqrt{I_{mn}}$) sums of $2^{-m}N$ consecutive output measurements y_k, $k = 2^{-m}Nn + 1, \ldots, 2^{-m}N(n+1)$, weighted by the corresponding to them spacings I_k.

Combining (6.7) and (6.9) for $m = \log_2 N$, we verify that (cf. (6.6)):

$$\hat{\alpha}_{\log_2 N-1,n} = \frac{1}{\sqrt{I_{\log_2 N-1,n}}} \sum_{k=2n+1}^{2n+2} I_k y_k$$

$$= \frac{I_{2n+1}}{\sqrt{I_{\log_2 N-1,n}}} y_{2n+2} + \frac{I_{2n+2}}{\sqrt{I_{\log_2 N-1,n}}} y_{2n+2}$$

$$= \frac{\sqrt{I_{\log_2 N,2n}}}{\sqrt{I_{\log_2 N-1,n}}} \hat{\alpha}_{\log_2 N,2n} + \frac{\sqrt{I_{\log_2 N,2n+1}}}{\sqrt{I_{\log_2 N-1,n}}} \hat{\alpha}_{\log_2 N,2n}, \tag{6.10}$$

and (6.8) holds. Note that in parallel to the computations of the empirical coefficients in (6.7), in each transform step also the intervals $I_{m-1,n}$ are simply computed as

$$I_{m-1,n} = I_{m,2n} + I_{m,2n+1}. \tag{6.11}$$

Analysis Phase

The fast implementation of the analysis phase of the EOS consists now of three steps:

- Sorting the set of the raw measurement pairs $\{(u_k, y_k)\}$ w.r.t. the ascending values of the input measurements (the ordered set will further be denoted by $\{(x_k, y_k)\}$)
- Applying the unbalanced Haar transform routine starting from the scale $m = \log_2 N - 1$ down to $m = K$ to the output measurements y_k prescaled by their corresponding factors $\sqrt{x_k - x_{k-1}}$
- Restoring the positions of the unbalanced scaling functions borders from the sequence of interval lengths I_{Kn} computed by (6.11)

Complexity

To sort the measurement pairs, we can apply, e.g., *heapsorting* algorithm which has a cost of $\mathcal{O}(N \log N)$ operations (cf., e.g., [89, Chap. 5.2.3]). The second step needs $\mathcal{O}(N)$ operations since—like in the classic fast wavelet transform—the number of constant time operations in (6.7),[2] for each scale m, is of order $\mathcal{O}(2^m)$, and to find the empirical coefficients at the scale $m = K \geq 0$, we also require

$$\sum_{m=K}^{\log_2 N-1} \mathcal{O}(2^m) = \mathcal{O}\left(2^{\log_2 N} - 2^K\right) = \mathcal{O}(N)$$

operations; cf. [96, Chap. 7.3.1]. The restoring procedure takes $\mathcal{O}(2^K) = \mathcal{O}\left(\sqrt[3]{N}\right)$ operations. The overall cost of the analysis phase is therefore of order $\mathcal{O}(N \log N)$—like in the ED algorithm.

Synthesis Phase

Evaluation of the estimate output can be implemented by the following formula:

$$\hat{\mu}(u) = \left. \sqrt{I_{Kn}^{-1}} \hat{\alpha}_{Kn} \right|_{n=\lfloor 2^K F_K(u) \rfloor},$$

where the operation $\lfloor 2^K F_K(u) \rfloor$ maps the input arguments u onto the translation index n of the corresponding empirical coefficients $\hat{\alpha}_{Kn}$.

The synthesis phase can be therefore computed in the same two-step way as in the ED algorithm:

- Determining the translation index n for a given argument u using the empirical distribution $F_K(u)$
- Evaluating the output value, i.e., scaling the selected coefficient $\hat{\alpha}_{Kn}$ by the associated factor $\sqrt{I_{Kn}^{-1}}$

Remark 6.3. The second step of this phase can also be performed in parallel with the last step of the analysis phase.

Complexity

The overall complexity of the synthesis phase equals the complexity of this phase in the ED algorithm, i.e., it is of order

---

[2]The lifting procedure (4.58) is also a constant time operation and can be used instead of (6.7) when the transform is to be executed in situ; cf. [24, 96].

$$\mathcal{O}\left(\log \sqrt[3]{N}\right) = \mathcal{O}\left(\log N\right),$$

and is larger than the constant one obtained for QOS and OS implementations.

6.2 Linear Algorithms Complexities

Taking into account the results of the linear algorithms complexity analyses, we infer the following corollary:

Corollary 6.1. *The computational complexities of the linear algorithms are log-linear w.r.t. the number of measurement data N.*

Table 6.1 Computational complexities of the linear algorithms

	Complexity		
	Analysis	Synthesis	Overall
QOS	$\mathcal{O}(N)$	$\mathcal{O}(1)$	$\mathcal{O}(N)$
OS	$\mathcal{O}(N \log N)$		$\mathcal{O}(N \log N)$
ED	$\mathcal{O}(N \log N)$	$\mathcal{O}(\log N)$	$\mathcal{O}(N \log N)$
EOS			

6.3 Nonlinear Algorithms

The implementation of the EZW nonlinear approximation scheme from section "EZW Approximation Scheme" is, in fact, an adaptive, data-driven, and empirical transformation of the set of the empirical coefficients $\hat{\beta}_{mn}$, $m = M, \ldots, K - 1$, $n = 0, \ldots, 2^m - 1$, produced by the analysis phase of the nonlinear algorithm, into the new set of coefficients $\hat{\beta}_{mn}$, $m = M, \ldots, K - 1$, $n \in Q_m$, which form q_M empirical cones of influence. As such, the implementation is independent of the type of the linear algorithm it is plugged-in, and in the subsequent chapters, we present the implementation without a reference to the particular identification algorithm.

The proposed EZW algorithm implementation follows the two-phase *analysis* and *synthesis* template and consists of:

- Execution of the fast wavelet transform routine computing the wavelet empirical coefficients $\hat{\beta}_{mn}$ at the scales $m = K - 1, \ldots, M$ along with the detection of the empirical cones of influence

- Removal of the unwanted (located outside the empirical cones) empirical wavelet coefficients $\hat{\beta}_{mn}$ during the fast inverse transform (to yield back the empirical scaling function coefficients $\hat{\alpha}_{Kn}$ at the scale K)

6.3.1 Implementation of the EZW Scheme

We assume that the number q_M of the empirical cones is selected according to the common to all nonlinear algorithms rule $q_M = \log_2 M$ (see, e.g., (5.21)).[3] Clearly, the actual number of jumps q and their locations is unknown, and the selected q_M *empirical cones* not necessary correspond to the actual jumps (in particular, when $q_M < q$, then not all jumps can be detected).

To find the empirical cones, we use the auxiliary vector

$$\hat{\beta} = \left[\hat{\beta}_0 \cdots \hat{\beta}_{2^{K-1}-1} \right], \tag{6.12}$$

storing the sums of the squares of the empirical coefficients $\hat{\beta}_{mn}$ grouped in all 2^{K-1} possible cones of influence.

Remark 6.4. The sums in (6.12) have the explicit form,

$$\hat{\beta}_p = \sum_{m=M}^{K-1} \hat{\beta}_{ml}^2, \text{ where } l = \left\lfloor 2^{M-m} p \right\rfloor, \tag{6.13}$$

for $p = 0, \ldots, 2^{K-1} - 1$ and can be computed in a separate routine; however, we will exploit the flexibility of the lifting transform scheme and will evaluate these sums in line with the forward wavelet transform routine.

We also need a list, L , which will store the q_M pairs $(\hat{\beta}_p, p)$ of the largest sums $\hat{\beta}_p$ and their indices p, indicating the locations of the empirical cones.

Remark 6.5. The proposed implementation of the EZW is based on the *heuristic* assumption that in all presented algorithms, the large empirical wavelet coefficients $\hat{\beta}_{mn}$ are located in the vicinity of each jump points of the nonlinearity (like their theoretical counterparts β_{mn}; cf. Fig. 4.2).

---

[3]Clearly, for a given number of measurements N, the scale M is fixed, and $q = \log_2 M$ is, from the implementation viewpoint, fixed as well.

Analysis Phase

Recall that the nonlinear algorithm starts when the empirical scaling function coefficients $\hat{\alpha}_{Kn}$ are already computed by the linear analysis part of the selected identification algorithm. Then, for each $m = K, \ldots, M+1$ and for each $n = 0, \ldots, 2^m - 1$, the following steps are performed:

- Execution of the standard forward transform step to evaluate the scaling function and wavelet empirical coefficients, $\hat{\alpha}_{m-1,n}$ and $\hat{\beta}_{m-1,n}$, from the scaling function empirical coefficients at the scale m (cf. (4.26) and (4.28))

$$\hat{\alpha}_{m-1,n} = \tfrac{1}{\sqrt{2}} \left(\hat{\alpha}_{m,2n} + \hat{\alpha}_{m,2n+1} \right) \tag{6.14}$$

$$\hat{\beta}_{m-1,n} = \tfrac{1}{\sqrt{2}} \left(\hat{\alpha}_{m,2n} - \hat{\alpha}_{m,2n+1} \right)$$

- Updating the appropriate sums $\hat{\beta}_p$, i.e., all sums with the index p such that

$$p = n \cdot 2^{m-M}, \ldots, (n + 1) \cdot 2^{m-M} - 1, \tag{6.15}$$

by adding to them the squared value of the newly computed empirical wavelet coefficient $\hat{\beta}_{mn}$.
- Inserting the pair $(\hat{\beta}_p, p)$ into the list L—if the value of the updated sum $\hat{\beta}_p$ is larger than the smallest currently stored there. If the pair is already present in the vector, then the value of $\hat{\beta}_p$ is only updated. Otherwise, it is replaced with the pair with the smallest sum.

At the end of this phase, we have at our disposal two sets of data: one with the vectors of empirical scaling function and wavelet coefficients $\hat{\alpha}_{Mn}$ and $\hat{\beta}_{mn}$, $m = M, \ldots, K - 1$, and the other, comprising the list L of q_M indices pointing out the largest sums, i.e., the selected *empirical cones of influence*.

Synthesis Phase

The synthesis phase is essentially a reverse of the analysis one and consists in removing (resetting) each of the wavelet empirical coefficients located outside the empirical cones of influence constructed prior to the inverse fast transform step. Namely, in this phase, one should, for each $m = M + 1, \ldots, K - 1$ and each $n = 0, \ldots, 2^m - 1$, perform the following steps:

- Check whether the currently processed empirical wavelet coefficient, $\hat{\beta}_{m-1,n}$, belongs to one of the empirical cones of influence, i.e., test whether the translation index n satisfies the condition

$$n = \lfloor 2^{m-K} p \rfloor,$$

for at least one of the q_M indices p from the list L. If not, reset it, i.e., set $\hat{\beta}_{m-1,n} = 0$.

- Then, execute of inverse FWT routine step (cf. (4.27) and (4.29))

$$\hat{\alpha}_{m,2n} = \tfrac{1}{\sqrt{2}} \left(\hat{\alpha}_{m-1,n} + \hat{\beta}_{m-1,n} \right) \tag{6.16}$$

$$\hat{\alpha}_{m,2n+1} = \tfrac{1}{\sqrt{2}} \left(\hat{\alpha}_{m-1,n} - \hat{\beta}_{m-1,n} \right).$$

After completion of the synthesis phase, we obtain the vector of new scaling function empirical coefficients $\hat{\alpha}_{Kn}$ which are ready to be scaled by the factor $2^{K/2}$ in order to produce (like in the linear algorithms) the final estimate of the nonlinearity.

Remark 6.6. The routine presented above needs to be adjusted to the NQOS algorithm quotient form, and in the analysis phase, the coefficients from the denominator, $\hat{a}_{mn}$ and $\hat{b}_{mn}$, should be computed analogically to the numerator coefficients $\hat{\alpha}_{mn}$ and $\hat{\beta}_{mn}$, using the formula in (6.14). In the synthesis phase, the empirical wavelet coefficients from numerator and denominator, $\hat{\beta}_{mn}$ and $\hat{b}_{mn}$, should also be jointly considered. That is, if for a given m and n, the coefficient $\hat{\beta}_{mn}$ is qualified for rejection, then so is its counterpart $\hat{b}_{mn}$ in the denominator; cf. Sect. 5.3. Finally, the estimate regressogram is produced as in the linear variant, by computing the quotients $\hat{\alpha}_{Kn} / \hat{a}_{Kn}$.

Remark 6.7. In the NEOS algorithm, the unbalanced transform steps in (4.58) and (4.59) should be used instead of those from (6.14) and (6.16). Furthermore, in order to obtain a proper regressogram, the resulting coefficients $\hat{\alpha}_{Kn}$ need to be scaled by their random factors $1/\sqrt{I_{Kn}}$ rather than by the constant one $2^{K/2}$.

6.3.2 Complexity of the EZW Scheme Implementation

In this chapter, we examine the complexity of the EZW analysis and synthesis routines to show that the overall complexity of the nonlinear plug-in implementation does not exceed the complexity of the preceding and following linear routines and hence that the orders of the computational complexity of both linear and nonlinear algorithms are equal.

Analysis Phase

For 2^K coefficients $\hat{\alpha}_{Kn}$, the standard FWT algorithm requires $\mathcal{O}\left(2^K\right)$ steps; however, in our one-pass implementation of the EZW scheme, in each transform step, we:

- Compute the new coefficients as in (6.14)—at the fixed cost $\mathcal{O}(1)$.
- Add the newly computed empirical coefficient $\hat{\beta}_{mn}$ to 2^{K-m} sums $\hat{\beta}_p$—this requires $\mathcal{O}(2^{K-m})$ extra operations; cf. (6.13) and (6.15).
- Check whether each new sum $\hat{\beta}_p$ should be inserted into the list L. If the list is implemented as a *min-heap* (see, e.g., [89, Chap. 5.2.3]) with all the initial sums equal to zero, then, in the worst case, such an insertion operation requires checking whether $\hat{\beta}_p$ is larger than the smallest sum in the heap, removing it and inserting the new pair $(\hat{\beta}_p, p)$. It costs $\mathcal{O}(1)$, $\mathcal{O}(\log q_M)$ and $\mathcal{O}(\log q_M)$, respectively.

Hence, for each $m = K, \ldots, M$, we perform $\mathcal{O}(2^K \log q_M)$ operations, and the resulting complexity of the analysis phase of the EZW implementation is of order $\mathcal{O}(2^K K \log M) = \mathcal{O}(N^{2/3} \log N \log \log N)$.

Synthesis Phase

The standard inverse FWT algorithm has the complexity $\mathcal{O}(2^K)$, i.e., the same as the forward one. For each coefficient, checking for the inclusion in the empirical influence cone takes $\mathcal{O}(\log q_M) = \mathcal{O}(\log \log M) = \mathcal{O}(\log \log \log N)$—if the list is implemented as a *min-heap*. Hence, the overall complexity of this phase of the EZW implementation is $\mathcal{O}(2^K \log q_M) = \mathcal{O}(N^{2/3} \log \log \log N)$, i.e., slightly smaller than in the analysis phase.[4]

6.4 Overall Complexity of the Nonlinear Algorithms

From the previous chapters, we know that the joint complexity of the analysis and synthesis phases of the linear algorithms is of the linear $\mathcal{O}(N)$ or the log-linear orders $\mathcal{O}(N \log N)$. Hence, in the view of the above analysis, the following corollary holds true (cf. Corollary 6.1 and Table 6.1).

Corollary 6.2. *The overall computational complexities of the proposed linear and nonlinear algorithms are the same and at most of order $\mathcal{O}(N \log N)$, i.e., they are log-linear w.r.t. the number of measurement data N; see Table 6.2.*

---

[4]In most practical situations, the iterated logarithm factor $\log \log \log N$ can be neglected.

Table 6.2 Computational complexities of the nonlinear algorithms

	Complexity		
	Analysis	Synthesis	Overall
NQOS	$\mathcal{O}(N)$	$\mathcal{O}(N^{2/3}\log\log\log N)$	$\mathcal{O}(N)$
NOS	$\mathcal{O}(N\log N)$	$\mathcal{O}(N^{2/3}\log\log\log N)$	$\mathcal{O}(N\log N)$
NED			
NEOS			

6.5 Computational Stability

Here, we shortly examine the computational stability of the implementations, that is, we check whether they produce bounded estimates for sets of bounded but random measurements (cf., e.g., [124, Prop. 3]). This kind of stability prevents from the overflow errors during computations (and can be seen as a counterpart of the classic *BIBO-stability* of linear systems).

Observe first that for both (linear and nonlinear) variants of the QOS algorithm, this property can be derived from the fact that they are bounded for any set of bounded measurements $\{(x_k, y_k)\}$; see Lemmas A.4 and A.6 in Appendix A.1 (note that the stability of the NQOS algorithm is strictly related to application of the EZW nonlinear approximation scheme and may not hold for, e.g., N-term approximation-based one).

One can also easily observe that the ED-type algorithms share this property as well: For the scaling function empirical coefficients in these algorithms, we have the inequality (cf. (5.34)):

$$|\hat{\alpha}_{Kn}| \leq \sum_{k=1}^{N} |y_k| \int_{\frac{k-1}{N}}^{\frac{k}{N}} \varphi_{Kn}(x)\,\mathrm{d}x \leq 2^{-K} N 2^{\frac{K}{2}} N^{-1} \cdot \max_{k=1,\dots,N}\{y_k\}$$

$$= 2^{-\frac{K}{2}} \max_{k=1,\dots,N} |y_k|,$$

which implies that, for each x, the estimate value is bounded for bounded output measurements (cf. (5.33))

$$|\hat{\mu}_K(x)| \leq \max_{k=1,\dots,N} |y_k|.$$

Recall now that for any measurements number N, the corresponding scale K is fixed. To establish the stability of the OS algorithm, consider the scaling function empirical coefficient of the OS algorithm, where we have that (cf. (5.23))

$$|\hat{\alpha}_{Kn}| \leq \sum_{k=1}^{N} |y_k| \int_{x_{k-1}}^{x_k} \varphi_{Kn}(x)\, dx$$

$$\leq 2^{\frac{K}{2}} N 2^{-K} \cdot \max_{k=1,\dots,N} |y_k| \max_{k=1,\dots,N} (x_k - x_{k-1})$$

$$\leq 2^{-\frac{K}{2}} N \cdot \max_{k=1,\dots,N} |y_k| ,$$

since the input signal is bounded, then the differences $x_k - x_{k-1}$ are all bounded as well and so is the estimated value (cf. (5.22))

$$|\hat{\mu}_K(x)| \leq N \cdot \max_{k=1,\dots,N} |y_k| .$$

Consider finally the EOS algorithm, where (cf. (5.44))

$$|\hat{\alpha}_{Kn}| \leq \sum_{k=1}^{N} |y_k| \int_{x_{k-1}}^{x_k} \varphi_{Kn}(x)\, dx \leq 2^{-K} N \frac{\sum_{k=1}^{2^{-K}N} I_k}{\sqrt{\sum_{k=1}^{2^{-K}N} I_k}} \cdot \max_{k=1,\dots,N} |y_k|$$

$$= 2^{-K} N \max_{k=1,\dots,N} |y_k| \sqrt{\sum_{k=1}^{2^{-K}N} I_k} ,$$

where the spacings I_k are bounded since the input signal is bounded. Thus, in this case (cf. (5.43)),

$$|\hat{\mu}_K(x)| \leq 2^{-K} N \cdot \max_{k=1,\dots,N} |y_k| .$$

Remark 6.8. The analysis for nonlinear variants of the OS, ED, and EOS algorithms can be performed using the analogous arguments with the help of their common representation in (5.52); cf. Sect. 5.10.2.

Chapter 7
Final Remarks

Abstract The features of the proposed algorithms are summarized. Their asymptotic properties and computational complexities are collected and compared. Finally, the performances of the linear and nonlinear algorithms in the considered nonparametric system identification tasks are discussed.

In this chapter we recapitulate the properties of the proposed Haar wavelet identification algorithms, highlight their advantages and shortcomings, and compare their asymptotic and computational properties.

7.1 Asymptotic Properties

Comparing the asymptotic properties of the proposed algorithms, we can point out their common advantages:

- They converge globally (in the MISE sense) to any (piecewise-)Lipschitz nonlinearities for any (piecewise-)Lipschitz input probability density functions.[1]
- For Lipschitz nonlinearities, the convergence rates of the linear algorithms are of order $\mathcal{O}\left(N^{-2/3}\right)$ and are optimal; cf. [58, 138, Chap. 4.2] and [72].[2]
- For discontinuous piecewise-Lipschitz nonlinearities, the linear algorithms achieve the guaranteed convergence rate $\mathcal{O}\left(N^{-1/2}\right)$, while the best-case convergence rates of the nonlinear algorithms are of order $\mathcal{O}\left(N^{-2/3}\right)$. The latter

---

[1] In fact, the class of piecewise-Lipschitz nonlinearities and input probability density functions can further be extended to the class of piecewise-Hölder functions (at the expense of more burdensome analysis, however); cf. e.g. [55, 110, 131, 132].

[2] Clearly, the convergence rates of the EOS-type algorithms are slowered by a logarithmic factor– and hence only near optimal.

P. Śliwiński, *Nonlinear System Identification by Haar Wavelets*, Lecture Notes in Statistics 210, DOI 10.1007/978-3-642-29396-2_7, © Springer-Verlag Berlin Heidelberg 2013

is of the same order as the rates of the linear algorithms for uniformly Lipschitz (i.e., continuous) nonlinearities. Such rates are not attainable by the algorithms based on any classic orthogonal series due to their global nature; cf. [16, 18, 29].

All these beneficial properties are achieved for any structure of linear (stable) dynamics in the system and for any second-order stationary external noise and make the algorithms robust against the system properties and the identification conditions; cf. [47, 55, 71, 72, 110]. Several weak points of the algorithms should however be also signalized:

– For discontinuous piecewise-Lipschitz nonlinearities, the worst-case convergence rate of the nonlinear algorithms $\mathcal{O}\left(N^{-1/3}\right)$ is slower than the rate $\mathcal{O}\left(N^{-1/2}\right)$ guaranteed for linear algorithms.
– The maximum rate of convergence is limited by the approximation abilities of Haar bases. In particular, that convergence rate cannot be faster that $\mathcal{O}\left(N^{-2/3}\right)$ even if the identified nonlinearities are smoother than Lipschitz (like, e.g., polynomials).

Remark 7.1. To overcome the first drawback, more efficient nonlinear approximation schemes need to be developed. To deal with the second deficiency and obtain faster convergence rates for (piecewise-)smoother nonlinearities, one should apply higher-order (and possessing more vanishing moments) compactly supported Daubechies wavelets (e.g., the wavelet bases constructed on interval as in [19]); see also [71, 131–133].

In addition to the above-mentioned common properties, there are also some algorithm specific attributes worth to be reported:

• Both QOS and OS linear algorithms maintain the asymptotic convergence rate $\mathcal{O}\left(N^{-2/3}\right)$ even when the nonlinearity is piecewise-Lipschitz but have jumps at dyadic points. Note that for such nonlinearities, application of the nonlinear variants of these algorithms cannot yield better convergence rate but, conversely, could rather deteriorate the estimate performance.
• All algorithms—except both variants of the quotient one—are not affected by the (lack of) smoothness of the input probability density function. Particularly, the nonlinear version of the QOS algorithms appears to be the least effective with the performance affected by discontinuities in the nonlinearity and in the density function.
• The scale of the estimates in ED and EOS algorithms self-adapts locally to the amplitude of the input density function. In this sense, their behavior is similar to *nearest neighbor algorithms* and to *optimal quantizers*; see Fig. 7.1 and cf. [58, Chap. 6] and [45, 94, 99, 136], respectively.

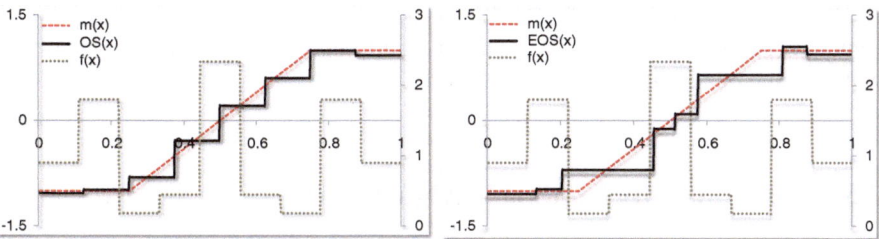

Fig. 7.1 The OS algorithm has the constant scale across the identification interval regardless of the density function shape. The EOS one self-adapts to the scale to the local measurements density (i.e., to the local height of the density function)

7.2   Computational Properties

The expedient computational properties are an important facet of our algorithms, and we deliberately have made a substantial effort to demonstrate (see the whole Chap. 6) that in spite of the random character of the processed measurement data, all the proposed implementations are computationally effective.[3] In particular, the number of operations performed in our implementations is of order $\mathcal{O}(N)$ or $\mathcal{O}(N \log N)$, i.e., it is linear or log-linear w.r.t. number of measurements N and stays in line with the complexity of the typical wavelet transform algorithms (cf. e.g. [96, Chap. 7.3]). Furthermore, despite of the extra overhead of the EZW nonlinear approximation scheme, the complexities of both linear and nonlinear variants of the algorithms are also the same (compare Tables 6.1 and 6.2). In the list below, we underline some of the algorithm-type specific aspects:

- The QOS algorithm is the only one which does not require sorting of the measurement sequence. However, due to its quotient form, the number of coefficients to be evaluated and stored is twice as much.
- The OS algorithm requires no explicit sorting, but the measurements set is implicitly sorted during its analysis phase. Note further that in spite of the form of the OS estimate empirical coefficients, they are computed without the use of numerical integration.
- Both QOS and OS algorithms require specific procedures for empirical coefficients evaluation. In contrast (for dyadic measurement numbers), the coefficients in the ED and EOS algorithms are computed directly from the measurements using the fast wavelet transform algorithms (the standard and unbalanced one, respectively).

---

[3]Such good properties were—to much extent—easier to accomplish due to the utmost simplicity of Haar functions and hierarchical construction of (both standard and unbalanced) orthogonal Haar bases which yields simple and fast transform routines.

Observe eventually that all algorithms enjoy the numerical stability property—a somehow surprising feature for the QOS algorithms which have the quotient form with random denominators, see Sect. 6.5.

7.3   Linear Versus Nonlinear Algorithms

Nonlinear algorithms have been proposed in attempt to improve the performance of identification in case when the nonlinearities are discontinuous and exploit the most distinguishing feature of the wavelet function—their local approximation property—which allows them to localize and isolate the separate jumps in nonlinearities and results in the fast, not available for other nonlocal orthogonal systems, convergence rates (see Table 5.1).

We emphasize however that the specific identification conditions imposed by the nonparametric assumptions and the presence of the dynamics-induced correlated system noise[4] make together the design of any nonlinear identification/estimation algorithm a much more arduous task than for other signal and image processing estimation problems (recall that the benchmark wavelet nonlinear algorithm—the thresholding routine by Donoho and Johnstone for signal denoising—was developed and formally examined for deterministic inputs and for normal white noise (see e.g., [33–35, 90]).[5]

Our nonlinear algorithms are clearly in their infancy. Nevertheless, they allow to formulate several observations which give a useful picture of nonlinear algorithms' potential and limitations in the nonparametric system identification area:

- Nonlinear algorithms faster (more aggressively) than their linear prototypes adapt the local estimate scale to the jumps' locations, for a given number of measurements N.
- Simultaneously, the nonadaptive linear algorithms are naturally more robust against the small signal-to-noise ratio (SNR).

Consequently, the data-dependent nature of nonlinear algorithms can neutralize their potential advantage over the linear ones as the SNR degrades.

---

[4]Also, in order to improve the performance of the EZW scheme, one can consider reduction of the system-induced noise by inverse filtering (as proposed in [104]).

[5]These conditions are—in our case—met only if the system is static, the external noise is normal, and the ED algorithm is used.

Appendix A
Technical Derivations

Abstract A detailed and more technical insight in the proposed algorithms is given. It starts from some preliminary lemmas. Then, the asymptotic properties, i.e., the convergence conditions and convergence rates of the algorithms, presented in the 'Identification Algorithms' chapter, are formally established.

In the remaining chapters, all the asymptotic results presented in Chap. 5 and describing the convergence conditions and the convergence rates are derived formally. The proofs follow a common scheme:

- The MISE error is decomposed and expressed in terms of the empirical coefficients errors.
- The convergence conditions are determined.
- The convergence rates for linear and nonlinear algorithms are derived.

The following three lemmas are exploited in the proofs of the algorithms' asymptotic properties. The first enables examination of the global properties of the QOS-type algorithms.

Lemma A.1 ([57, Chaps. 6.8 and Eq. (C.1)]). *The following inequality holds*

$$\left| \frac{y_k}{x_k} - \frac{b}{a} \right| \leq \left| \frac{y_k}{x_k} \right| \left| \frac{x_k - a}{a} \right| + \left| \frac{y_k - b}{b} \right|.$$

The other two are extensively used in the proofs of the remaining types of algorithms (recall that $\xi_k = \sum_{i=0, i \neq d}^{\infty} \lambda_i m (u_{k-i}) - b_d$ with $b_d = Em (u_1) \sum_{i=0, i \neq d}^{\infty} \lambda_i$ and thus $E\xi_k = 0$; cf. (2.2)).[1]

---

[1]In the proofs, we will use the generic symbol c to denote any positive constant independent of the accompanying factors.

P. Śliwiński, *Nonlinear System Identification by Haar Wavelets*, Lecture Notes in Statistics 210, DOI 10.1007/978-3-642-29396-2,
© Springer-Verlag Berlin Heidelberg 2013

Lemma A.2 (cf. [57, L. 7.1]). *Let $\mu(x)$ be piecewise-Lipschitz in the interval* $[0, 1]$. *Then, for any nonnegative Borel function $\zeta(x)$ and for any i, j, k, l, all different,*

$$E\left\{\xi_i^2 \middle| \zeta(x_i, x_j)\right\} \le cE\left\{\zeta(x_i, x_j)\right\}, \tag{A.1}$$

and

$$E\left\{|\xi_i \xi_i| \, \zeta(x_i, x_j, x_k, x_l)\right\} \le cN^{-1} E\left\{\zeta(x_i, x_j, x_k, x_l)\right\}. \tag{A.2}$$

Lemma A.3 ([57, App. C.4.2]). *Let $\{x_k\}$, $k = 1, \ldots, N$, be the set of ordered independent random variables with a density function $f(x)$ such that $c \le f(x) \le C$, for some $C \ge c > 0$, and all $x \in [0, 1]$. Then, for all $i, j, k = 1, \ldots, N + 1$ (taking $x_0 = 0$ and $x_{N+1} = 1$), we have for $p = 1, 2, \ldots$, that*

$$E(x_k - x_{k-1})^p \le cN^{-p}$$

and, in particular

$$E(x_i - x_{i-1})(x_j - x_{j-1}) \le cN^{-2}. \tag{A.3}$$

A.1   QOS Algorithms

The MISE error used to examine the QOS estimate $\hat{\mu}_K(x)$ presented in Sect. 5.2 is defined in a standard way as

$$\text{MISE } \hat{\mu}_K = E \int_0^1 [\hat{\mu}_K(x) - \mu(x)]^2 \, dx.$$

The analysis of the error requires some preliminary steps. In the analysis, we will exploit the explicit quotient form of the estimate. Denote by $\hat{g}_K(x)$ the numerator of the estimate and by $\hat{f}_K(x)$ its denominator and recall that

$$\hat{\alpha}_{Kn} = \frac{1}{N} \sum_{k=1}^N y_k \varphi_{Kn}(x_k) \text{ and } \hat{a}_{Kn} = \frac{1}{N} \sum_{k=1}^N \varphi_{Kn}(x_k).$$

Since ξ_1 and z_1 are both zero-mean and independent of $\mu(x_1)$ and $\varphi_{Kn}(x_1)$, we get

$$E\hat{\alpha}_{Kn} = Ey_1 \varphi_{Kn}(x_1) = E\left[\mu(x_1) + \xi_1 + z_1\right] \varphi_{Kn}(x_1)$$

$$= E\mu(x_1) \varphi_{Kn}(x_1)$$

$$= \int_0^1 \mu(x) f(x) \varphi_{Kn}(x) \, dx = \int_0^1 g(x) \varphi_{Kn}(x) \, dx,$$

where $g(x) = \mu(x) f(x)$ is a product of the identified nonlinearity $\mu(x)$ and the input signal probability density function $f(x)$. This means that $\hat{\alpha}_{Kn}$'s are unbiased estimates of the expansion coefficients α_{Kn}'s of this product. Similarly, observing that

$$E\hat{a}_{Kn} = E\varphi_{Kn}(x_1) = \int_0^1 f(x)\,\varphi_{Kn}(x)\,\mathrm{d}x,$$

we infer that the coefficients $\hat{a}_{Kn}$ are unbiased estimates of $f(x)$.

Because of unbiasedness of the empirical coefficients $\hat{\alpha}_{Kn}$ and $\hat{a}_{Kn}$, for each scale K, we have that $E\hat{g}_K(x) = g_K(x)$ and $E\hat{f}_K(x) = f_K(x)$. That is, for each K, the numerator estimates the approximation $g_K(x)$ of the product $g(x)$. Since the denominator estimates the approximation $f_K(x)$ of the density function $f(x)$, then one can expect that the quotient $\hat{\mu}_K(x)$ estimates an approximation $\mu_K(x)$ of the nonlinearity $\mu(x)$.

A.1.1 Convergence

To show the MISE error convergence, we will use an additional assumption that *the external noise signal is bounded*, that is, we assume that there exists an unknown constant c such that $\max_k \{|z_k|\} \le c$ (and hence $|y_k| \le c$ for some c and all $k = 0, 1, \ldots$); cf. Sect. 5.2. The following lemma holds by virtue of this assumption.

Lemma A.4. *The estimate in QOS algorithm is bounded for any set of bounded measurements.*

Proof. The lemma says that for a bounded input, the QOS estimate yields bounded output. It is rather intuitive observing that, in fact, the estimate computes local average in the intervals determined by the supports of the estimate scaling functions; cf. (5.9). To prove it formally, we exploit the equivalent kernel representation in (5.10)

$$\left|\frac{\hat{g}_K(x)}{\hat{f}_K(x)}\right| = \left|\frac{\sum_{n=0}^{2^K-1}\hat{\alpha}_{Kn}\varphi_{Kn}(x)}{\sum_{n=0}^{2^K-1}\hat{a}_{Kn}\varphi_{Kn}(x)}\right| = \frac{\sum_{k=1}^N |y_k|\,\phi_K(x_k,x)}{\sum_{k=1}^N \phi_K(x_k,x)} \le \max_k |y_k| \quad \text{(A.4)}$$

where $\phi_K(v,x)$ is the Haar reproducing kernel, i.e., a nonnegative window kernel function; cf. Sect. 4.1.3 and [147, Chap. 3]. ∎

Recalling that, by Assumption 1, $f(x)$ is strictly positive in the identification interval $[0, 1]$, i.e., it holds that $f(x) \ge c > 0$, we can now use the inequality (cf. Lemma A.1):

$$\left[\frac{\hat{g}_K(x)}{\hat{f}_K(x)} - \frac{g(x)}{f(x)} \right]^2 \le 2 \left[\frac{\hat{g}_K(x)}{\hat{f}_K(x)} \right]^2 \frac{1}{f^2(x)} \left[\hat{g}_K(x) - g(x) \right]^2 \qquad (A.5)$$

$$+ 2 \frac{1}{f^2(x)} \left[\hat{f}_K(x) - f(x) \right]^2$$

to bound the MISE error of the estimate $\hat{\mu}_K(x)$ by the following MISE errors of the estimate numerator, $\hat{g}_K(x)$, and the denominator, $\hat{f}_K(x)$:

$$\text{MISE } \hat{\mu}_K \le 2cE \int_0^1 [\hat{g}_K(x) - g(x)]^2 \, dx + 2cE \int_0^1 \left[\hat{f}_K(x) - f(x) \right]^2 dx$$

$$= 2c \, \text{MISE } \hat{g}_K + 2c \, \text{MISE } \hat{f}_K.$$

In other words, if both the numerator $\hat{g}(x)$ and the denominator $\hat{f}(x)$ converge in the MISE error sense to the product $g(x)$ and to the density $f(x)$, respectively, then the whole estimate $\hat{\mu}(x)$ converges to the nonlinearity $\mu(x)$.

To examine the MISE error of the numerator $\hat{g}(x)$, we use its standard decomposition into the integrated variance and integrated squared bias error components

$$\text{MISE } \hat{g}_K = E \int_0^1 [\hat{g}_K(x) - g(x)]^2 \, dx$$

$$= E \int_0^1 [\hat{g}(x) - g_K(x) + g_K(x) - g(x)]^2 \, dx$$

$$= E \int_0^1 [\hat{g}_K(x) - g_K(x)]^2 \, dx + E \int_0^1 [g_K(x) - g(x)]^2 \, dx$$

$$= \text{IV } \hat{g}_K + \text{ISB } \hat{g}_K.$$

Both errors can be further decomposed and expressed in terms of the variance and squared bias errors of the empirical coefficients.

$$\text{ISB } \hat{g}_K = \int_0^1 \left\{ \sum_{n=0}^{2^K-1} \alpha_{Kn} \varphi_{Kn}(x) \right.$$

$$\left. - \left[\sum_{n=0}^{2^K-1} \alpha_{Kn} \varphi_{Kn}(x) + \sum_{m=K}^{\infty} \sum_{n=0}^{2^m-1} \beta_{mn} \psi_{mn}(x) \right] \right\}^2 dx$$

$$= \int_0^1 \left[\sum_{m=K}^{\infty} \sum_{n=0}^{2^K-1} \beta_{mn} \psi_{mn}(x) \right]^2 dx.$$

Applying wavelets orthonormality argument (see (4.20)) leads to the observation that

$$\text{ISB } \hat{g}_K = \sum_{m=K}^{\infty} \sum_{n=0}^{2^K-1} \sum_{m'=K}^{\infty} \sum_{n'=0}^{2^K-1} \beta_{mn} \beta_{m'n'} \int_0^1 \psi_{mn}(x) \psi_{m'n'}(x) \, dx$$

$$= \sum_{m=K}^{\infty} \sum_{n=0}^{2^K-1} \beta_{mn}^2 = \text{ISE } g_K,$$

i.e., that the integrated squared bias error is of pure deterministic nature and is equal to the sum of squared wavelet expansion coefficients not present in the approximation $g_K(x)$. Clearly, by virtue of the Parseval's identity,

$$\text{ISB } \hat{g}_K = \sum_{m=K}^{\infty} \sum_{n=0}^{2^K-1} \beta_{mn}^2 \to 0 \text{ as } K \to \infty. \tag{A.6}$$

We are now passing to the variance error analysis. We have

$$\text{IV } \hat{g}_K = E \int_0^1 \left[\sum_{n=0}^{2^K-1} \hat{\alpha}_{Kn} \varphi_{Kn}(x) - \sum_{n=0}^{2^K-1} \alpha_{Kn} \varphi_{Kn}(x) \right]^2 dx.$$

Using now the scaling function orthonormality (see (4.6)), we get that

$$\text{IV } \hat{g}_K = E \sum_{n=0}^{2^K-1} \sum_{n'=0}^{2^K-1} [\hat{\alpha}_{Kn} - \alpha_{Kn}][\hat{\alpha}_{Kn'} - \alpha_{Kn'}] \int_0^1 \varphi_{Kn}(x) \varphi_{Kn'}(x) \, dx$$

$$= \sum_{n=0}^{2^K-1} E [\hat{\alpha}_{Kn} - \alpha_{Kn}]^2 = \sum_{n=0}^{2^K-1} \text{var } \hat{\alpha}_{Kn}. \tag{A.7}$$

That is, that the integrated variance of the estimate $\hat{g}_K(x)$ is just a sum of the variances of its empirical coefficients $\hat{\alpha}_{Kn}$. For these variances, the following lemma holds.

Lemma A.5. *The variance of the empirical scaling function coefficients, $\hat{\alpha}_{Kn}$, vanishes with growing number of measurements, N, and satisfies the following inequality:*

$$\text{var } \hat{\alpha}_{Kn} = c N^{-1}. \tag{A.8}$$

Proof. The proof is actually a Haar wavelet version of the proofs given in, e.g., [57, Chap. 6.9] for 'classic' quotient orthogonal series estimates (cf. [110]; the versions for compactly supported wavelets can be found in, e.g., [71, 72]).

It exploits the standard decomposition of the variance of the sum of correlated random variables:

$$\mathrm{var}\,\hat{\alpha}_{Kn} = \mathrm{var}\left\{\frac{1}{N}\sum_{k=1}^{N} y_k \varphi_{Kn}(x_k)\right\}$$

$$= \frac{1}{N^2}\sum_{k=1}^{N}\mathrm{var}\left\{y_k\varphi_{Kn}(x_k)\right\}$$

$$+\frac{1}{N^2}\sum_{\substack{i=1\ j=1 \\ j\neq i}}^{N}\sum^{N}\mathrm{cov}\left\{y_i\varphi_{Kn}(x_i),y_j\varphi_{Kn}(x_j)\right\}.$$

Starting with a variance part (we use $\varphi_k = \varphi_{Kn}(x_k)$ for shortness), we get

$$\frac{1}{N^2}\sum_{k=1}^{N}\mathrm{var}\left\{y_k\varphi_k\right\} = \frac{1}{N^2}\sum_{k=1}^{N}\mathrm{var}\left\{(\mu_k + \xi_k + z_k)\cdot\varphi_k\right\},$$

where, because for each k, the random variables φ_k, ξ_k, and z_k are independent and zero-mean, we get

$$\mathrm{var}\left\{\mu_k\cdot\varphi_k + \xi_k\cdot\varphi_k + z_k\cdot\varphi_k\right\} \leq E\mu_k^2\cdot\varphi_k^2 + E\xi_k^2\cdot E\varphi_k^2 + Ez_k^2\cdot E\varphi_k^2$$

$$= E\mu_1^2\varphi_1^2 + \left[E\xi_1^2 + Ez_1^2\right]\cdot E\varphi_1^2$$

$$\leq c,$$

for some c independent of K and n (as $\mu(x)$ is bounded and $E\varphi_1^2 \leq 1$).

The analysis of the covariance part seems to be much more intricate, and we skip (for clarity) the external noise z_k.

Recall now that using a stationarity argument, we have that $\mathrm{cov}\left\{y_i\varphi_i, y_j\varphi_j\right\} = \mathrm{cov}\left\{y_j\varphi_j, y_i\varphi_i\right\}$. Hence,

$$\sum_{\substack{i=1\ j=1 \\ j\neq i}}^{N}\sum^{N}\mathrm{cov}\left\{y_i\varphi_i, y_j\varphi_j\right\} = 2\sum_{i=1}^{N}\sum_{j=i+1}^{N}\mathrm{cov}\left\{y_i\varphi_i, y_j\varphi_j\right\}.$$

Exploiting the standard decomposition of the covariance, we get

$$\mathrm{cov}\left\{y_i\varphi_i, y_j\varphi_j\right\} = E\left\{y_i\varphi_i y_j\varphi_j\right\} - E\left\{y_i\varphi_i\right\}E\left\{y_j\varphi_j\right\},$$

where for the latter expectations, we clearly have (since μ_i and ξ_j are independent for $i \leq j$ and $E\xi_i = E\xi_1 = 0$) that

$$E\{y_i\varphi_i\} = E\{\mu_i\varphi_i\} + E\{\xi_i\varphi_i\} = E\{\mu_i\varphi_i\},$$

and similarly $E\{y_j\varphi_j\} = E\{\mu_j\varphi_j\}$. Next, observe that

$$
\begin{aligned}
E\{y_i\varphi_i y_j\varphi_j\} &= E\{(\mu_i + \xi_i)(\mu_j + \xi_j)\varphi_i\varphi_j\} \\
&= E\{\mu_i\mu_j\varphi_i\varphi_j + \mu_i\xi_j\varphi_i\varphi_j\} + E\{\xi_i\mu_j\varphi_i\varphi_j + \xi_i\xi_j\varphi_i\varphi_j\} \\
&= E\{\mu_i\mu_j\varphi_i\varphi_j\} + E\{\mu_i\xi_j\varphi_i\varphi_j\} \\
&\quad + E\{\xi_i\mu_j\varphi_i\varphi_j\} + E\{\xi_i\xi_j\varphi_i\varphi_j\} \\
&= E\{\mu_i\varphi_i\}E\{\mu_j\varphi_j\} + E\{\mu_i\xi_j\varphi_i\varphi_j\} \\
&\quad + E\{\xi_i\mu_j\varphi_i\varphi_j\} + E\{\xi_i\xi_j\varphi_i\varphi_j\},
\end{aligned}
$$

and hence

$$
\begin{aligned}
\operatorname{cov}\{y_i\varphi_i, y_j\varphi_j\} &= E\{\xi_j\mu_i\varphi_i\}E\{\varphi_j\} + E\{\varphi_i\}E\{\xi_i\}E\{\mu_j\varphi_j\} \\
&\quad + E\{\xi_i\xi_j\varphi_i\varphi_j\} \\
&= E\{\xi_j\mu_i\varphi_i\}E\{\varphi_j\} + E\{\xi_i\xi_j\varphi_i\varphi_j\}.
\end{aligned}
$$

Thus,

$$
\begin{aligned}
2\sum_{i=1}^{N}\sum_{j=i+1}^{N}\operatorname{cov}\{y_i\varphi_i, y_j\varphi_j\} &= 2\sum_{i=1}^{N}\sum_{j=i+1}^{N} E\{\varphi_j\}E\{\mu_i\varphi_i\xi_j\} \\
&\quad + 2\sum_{i=1}^{N}\sum_{j=i+1}^{N} E\{\varphi_j\}E\{\varphi_i\xi_i\xi_j\}.
\end{aligned}
$$

Recalling now that $\xi_j = \sum_{i=1}^{\infty}\lambda_i\zeta_{j-i}$, where $\zeta_{j-i} = m_{j-i} - Em_1$ (cf. Proposition 2.1), we get

$$E\{\mu_i\varphi_i\xi_j\} = E\left\{\mu_i\varphi_i \sum_{k=1}^{\infty}\zeta_{j-k}\lambda_k\right\} = E\{\mu_i\varphi_i\zeta_i\lambda_{j-i}\} = \lambda_{j-i}E\{\varphi_i\mu_i\zeta_i\}$$

and

$$E\left\{\varphi_i \xi_i \xi_j\right\} = E\left\{\varphi_i \sum_{k=1}^{\infty} \zeta_{i-k}\lambda_k \sum_{k=1}^{\infty} \zeta_{j-k}\lambda_k\right\}$$

$$= E\left\{\sum_{k=1}^{\infty} \varphi_i \zeta_{i-k}^2 \lambda_k \lambda_{j+k-i}\right\}$$

$$= \sum_{k=1}^{\infty} E\varphi_i E\zeta_{i-k}^2 \lambda_k \lambda_{j+k-i}.$$

Finally, noting that $|\mu(x)|, |\xi(x)| < c$ for all x (since the nonlinearity $m(x)$ is bounded) and that

$$E\left|\varphi_{Kn}(x_1)\right| = 2^{\frac{K}{2}} \int_{\mathrm{supp}\,\varphi_{Kn}} f(x)\,\mathrm{d}x \le 2^{-\frac{K}{2}} \le 1,\ \text{for all } K \ge 0,$$

we obtain

$$2\sum_{i=1}^{N}\sum_{j=i+1}^{N} \left|\mathrm{cov}\left\{y_i\varphi_i, y_j\varphi_j\right\}\right| \le 2\sum_{i=1}^{N}\sum_{j=i+1}^{N} E\varphi_j \left|\lambda_{j-i}\right| E\left\{\varphi_i \mu_i \zeta_i\right\}$$

$$+2\sum_{i=1}^{N}\sum_{j=i+1}^{N}\sum_{k=1}^{\infty} E\varphi_j E\varphi_i E\zeta_{i-k}^2 \left|\lambda_k\right|\left|\lambda_{j+k-i}\right|$$

$$\le 2c\sum_{i=1}^{N}\sum_{j=i+1}^{N}\left[\left|\lambda_{j-i}\right| E\left|\varphi_j\right| E\left|\varphi_i\right| + \sum_{k=1}^{\infty} E\left|\varphi_j\right| E\left|\varphi_i\right|\left|\lambda_k\right|\left|\lambda_{j+k-i}\right|\right]$$

$$\le 2c\sum_{i=1}^{N}\sum_{j=i+1}^{N}\left[\left|\lambda_{j-i}\right| + \sum_{k=1}^{\infty}\left|\lambda_k\right|\left|\lambda_{j-i+k}\right|\right]$$

$$= 2cN\sum_{k=1}^{N}\left(1 - \frac{k}{N}\right)\left[\left|\lambda_k\right| + \sum_{l=1}^{\infty}\left|\lambda_l\right|\left|\lambda_{k+l}\right|\right]$$

and

$$2\sum_{i=1}^{N}\sum_{j=i+1}^{N} \mathrm{cov}\left\{y_i\varphi_i, y_j\varphi_j\right\} \le 2cN\sum_{k=1}^{N}\left[\left|\lambda_k\right| + \sum_{l=1}^{\infty}\left|\lambda_l\right|\left|\lambda_{k+l}\right|\right]$$

$$\le N\cdot c.$$

Hence, for each K and n, we get that

$$\mathrm{var}\,\hat{\alpha}_{Kn} \le \frac{c}{N} + \frac{Nc}{N^2} \le cN^{-1}.$$

■

Combining now the last lemma with the formula in (A.7), we get that the integrated variance of the estimate $\hat{g}_K(x)$ depends on both the scale of the estimate and the number of measurements:

$$\text{IV}\,\hat{g}_K \leq c\frac{2^K}{N}. \tag{A.9}$$

One can easily ascertain that:

- For a fixed scale K, the variance vanishes as number of measurements, N, tends to infinity.
- For a fixed number of measurements, the variance grows (exponentially) with a growing scale K.

Combining (A.6) and (A.9), we obtain that

$$\text{MISE}\,\hat{g}_K \leq \sum_{m=K}^{\infty} \sum_{n=0}^{2^K-1} \beta_{mn}^2 + c\frac{2^K}{N}.$$

To make the MISE error vanishing, we need both the approximation and variance terms to vanish. However, with the scale factor K growing to infinity, the approximation error decreases, but the variance increases. Thus, the rate of the growth of the scale K needs to be selected in accordance with the number of measurements N and slow enough to allow the variance part to vanish with the growing number of measurements N too. Clearly, taking $K = \eta \log_2 N$ with any $0 < \eta < 1$, we get the MISE convergence since then $K \to \infty$ (i.e., ISB $\hat{g}_K \to 0$) and $2^K/N \to 0$ (i.e., IV $\hat{g}_K \to 0$) as $N \to \infty$.

In the similar way, one can show that the same conditions need to be satisfied to make the denominator estimate $\hat{f}_K$ converges in the MISE error sense. In this way, by virtue of the convergences of the numerator and the denominator and the inequality (A.5), we have shown the convergence of the whole estimate and proved the Theorem 5.1.

A.1.2 Convergence Rates

We know from Lemma 4.4 that the approximation error ISB depends on smoothness of the approximated function. In contrast, from the bounds in (A.8) and (A.9), we conclude that the variance error is smoothness independent. Examining the convergence rates, we first consider the case when both the nonlinearity $\mu(x)$ and the input probability density function $f(x)$ are uniformly smooth (Lipschitz).

Lipschitz Case: Assume that $\mu(x)$ and $f(x)$ are Lipschitz. Then, the product function $g(x)$ is also Lipschitz, since

$$|g(x) - g(v)| = |\mu(x) f(x) - \mu(v) f(x) + \mu(v) f(x) - \mu(v) f(v)|$$
$$\leq |f(x)| |\mu(x) - \mu(v)| + |\mu(v)| |f(x) - f(v)|$$
$$\leq c|x - v| + c|x - v|.$$

The approximation error in (A.6) satisfies the following inequality:

$$\text{ISB } \hat{g}_K = \text{ISE } g_K \leq c2^{-2K},$$

and for the MISE error, we get that

$$\text{MISE } \hat{g}_K \leq c \left(2^{-2K} + \frac{2^K}{N} \right) = c \left(2^{-2\eta \log_2 N} + \frac{2^{\eta \log_2 N}}{N} \right).$$

From the proof of Theorem 5.1, we already know that the convergence holds for $K = \eta \log_2 N$ with any $0 < \eta < 1$, and now we will try to select η so that the convergence rate is the fastest possible. Observing again the opposite behavior of the MISE error components with respect to the estimate scale K, we conclude that the asymptotically optimal convergence rate will be obtained when both the variance and the bias errors vanish with the same rate (otherwise, the overall MISE error would vanish with the rate of the slower component). To find such η, we need to solve the equation

$$2^{-2\eta \log_2 N} = \frac{2^{\eta \log_2 N}}{N},$$

which yields that $\eta = 1/3$. That is, taking as the scale selection rule

$$K = \tfrac{1}{3} \log_2 N$$

we obtain the following convergence rate of the estimate numerator $\hat{g}_K(x)$:

$$\text{MISE } \hat{g}_K \leq cN^{-2/3}.$$

Applying the above arguments to the estimate denominator $\hat{f}_K(x)$, we get

$$\text{MISE } \hat{f}_K \leq cN^{-2/3},$$

which together with the inequality (A.5) yields the convergence rate of the estimate $\hat{\mu}_K(x)$ claimed in the Theorem 5.2.

Piecewise-Lipschitz Case: Assume now that both the nonlinearity and the density function are piecewise-smooth. From Lemma 4.5, we derive the bound for the approximation error of the product $g(x)$

$$\text{ISB } \hat{g}_K = \text{ISE } g_K \leq c2^{-K},$$

that is (in accordance with intuition), the approximation error vanishes slower than for uniformly smooth nonlinearities. For the MISE error, it holds that

$$\text{MISE } \hat{g}_K \leq c \left(2^{-K} + \frac{2^K}{N} \right),$$

and—clearly—the scale selection rule which assures the same convergence rate of both components is as in (5.14), i.e.,

$$K = \tfrac{1}{2} \log_2 N.$$

The resulting MISE error of the estimate $\hat{g}_K(x)$ is thus

$$\text{MISE } \hat{g}_K \leq c N^{-1/2}.$$

Again, after applying the similar arguments to the estimate $\hat{f}_K(x)$, we obtain that

$$\text{MISE } \hat{g}_K \leq c N^{-1/2},$$

Inserting both inequalities into (A.5) ends the proof of Theorem 5.6.

To verify Corollary 5.1, it suffices to observe that for the scale selection rule in (5.12) and piecewise-Lipschitz nonlinearity (or density function), the variance error bound remains the same, but the approximation error components decay slower as

$$\text{ISB } \hat{g}_K, \text{ISB } \hat{f}_K \leq c 2^{-\frac{1}{3} \log_2 N} = c N^{-1/3},$$

and are responsible for the slower convergence rate of the estimate, since now

$$\text{MISE } \hat{\mu}_K \leq c \left(N^{-1/3} + N^{-2/3} \right) \leq c N^{-2/3}.$$

Binary Rational Case: The slower (asymptotic) convergence rate for piecewise-continuous nonlinearities occurs only when the jump points are not located at the *binary rationals*. To see this consider the function $g(x)$ with jumps at binary rationals and Lipschitz between jumps. The following decomposition holds

$$g(x) = g'(x) + g''(x)$$

where $g'(x)$ is a Lipschitz-continuous function and where $g''(x)$ is a piecewise-constant function with jumps at binary rationals, $2^{-K'} n$, for some K' (thus, it is a *trivial* function belonging to the space $V_{K'}$; cf. Remark 4.6). The error of approximation of such function by the approximant $g_K(x)$ can be decomposed into the 'smooth' and 'piecewise-constant' terms:

$$\text{ISE}\, g_K(x) = \int_0^1 \left\{ g_K'(x) + g_K''(x) - \left[g'(x) + g''(x) \right] \right\}^2 dx$$

$$\leq 2 \int_0^1 \left[g_K'(x) - g'(x) \right]^2 dx + 2 \int_0^1 \left[g_K''(x) - g''(x) \right]^2 dx.$$

For all scales, $K \geq K'$, the second term is zero. This is because the empirical coefficients $\hat{\alpha}_{Kn}$ are unbiased, and

$$E\hat{\alpha}_{Kn} = \alpha_{Kn} = \int_0^1 \left[g'(x) + g''(x) \right] \varphi_{Kn}(x)\, dx$$

$$= \int_0^1 g'(x)\, \varphi_{Kn}(x)\, dx + \int_0^1 g''(x)\, \varphi_{Kn}(x)\, dx$$

$$= \alpha'_{Kn} + \alpha''_{Kn}, \text{ for all } n = 0, \ldots, 2^K - 1,$$

where α'_{Kn} and α''_{Kn} are clearly the scaling function coefficients of $g'(x)$ and $g''(x)$, respectively. Hence,

$$E\hat{g}''(x) = g_K''(x) = g''(x) \in V_K,$$

for all $K \geq K'$, and the overall approximation error is equal to the error of the smooth part; cf. Remark 4.6

$$\text{ISE}\, g_K(x) = \text{ISE}\, g_K'(x) \leq c2^{-2K}.$$

Observing that the same decomposition holds for the density $f(x)$, we can conclude the proof.

A.1.3 Nonlinear Algorithm

In order to show the MISE error convergence (and the convergence rates) of the nonlinear QOS algorithm, we need the following lemma; cf. also Sect. 4.2.4.

Lemma A.6. *The nonlinear QOS algorithm with EZW scheme preserves the boundedness property of its linear version.*

Proof. The proof exploits the equivalent kernel representation employing the standard Haar kernel function $\vartheta_K(x, v)$, in which the kernel scale depends on the argument x.

Assume for simplicity that the wavelet coefficients are in pth cone of influence. Since by design, our EZW algorithm collects all the empirical wavelet coefficients at the scales $m = M, \ldots, K - 1$ which belong to the cone, then for all x in the interval $[2^{-(K-1)}p, 2^{-(K-1)}(p+1))$, the equivalent kernel has the scale K, in the wider interval $[2^{-(K-2)}p, 2^{-(K-2)}(p+1))$ its scale equals $K-1$, and so on down to the scale M. We can thus represent the nonlinear estimate numerator as

$$\hat{g}_{MK}(x) = \sum_{k=1}^{N} y_k \phi_{K(x)}(x, x_k).$$
(A.10)

where $K(x)$ is some (data dependent) function of x. Since in the estimate denominator only the coefficients $\hat{b}_{mn}$ with the same indices m and n are pruned, then for the same $K(x)$, we get that

$$\hat{f}_{MK}(x) = \sum_{k=1}^{N} \phi_{K(x)}(x, x_k),$$
(A.11)

and therefore, we can show boundedness of the whole nonlinear estimate, $\hat{\mu}_K(x) = \hat{g}_K(x) / \hat{f}_K(x)$, using the same arguments as in the linear case; cf. (A.4). ■

The next lemma, stating that all the empirical wavelet coefficients have the same variance bound, has been pivotal for the nonlinear algorithm construction.

Lemma A.7. *The variance of the empirical wavelet coefficients, $\hat{\beta}_{Kn}$, vanishes with growing number of measurements, N, and satisfies the following inequality*

$$\operatorname{var} \hat{\beta}_{Kn} = cN^{-1}$$
(A.12)

for some $c_{\operatorname{var}\hat{\beta}} > 0$, independent of the scale and translation factors m and n; cf. [63].

Proof. The of the lemma is tantamount to the proof of Lemma A.5 (cf., e.g., [71, 72]). ■

The lemma says that asymptotically the empirical wavelet coefficients are indistinguishable from the variance error viewpoint and therefore only their approximation properties remain discriminative (in consequence, it allows for a direct implementation of the nonlinear approximation techniques from Sect. 4.1.13 in the nonlinear identification algorithms from Chap. 5).

Selection Rule for q_M: The number of the empirical cones constructed in the algorithm q_M needs to grow with the scale M to assure that asymptotically it exceeds an arbitrarily large number of discontinuities in $\mu(x)$ and $f(x)$). The restriction (A.13) imposed on the rate of that growth guarantees that the extra variance, induced by $q_M(K-M)$ additional empirical wavelet coefficients of the nonlinear part, is at most of the same order as the variance of the linear base, that is, we can take any q_M which satisfies the inequality

$$q_M(K-M) \leq 2^M.$$

Recalling that $K = 2M$, we have

$$q_M \leq 2^M M^{-1},$$

and, in particular, the rate in (5.21) is valid.

Convergence: Using an orthonormality argument, we find that the integrated variance of the nonlinear estimate $\hat{g}_{MK}(x)$ equals to the sum of variances of all empirical coefficients

$$
\begin{aligned}
\text{IV}\, \hat{g}_{MK} &= E \int_0^1 [\hat{g}_{MK}(x) - g_{MK}(x)]\, dx \\
&= \sum_{n=0}^{2^M-1} E\, (\hat{\alpha}_{Mn} - \alpha_{Mn})^2 + \sum_{m=M}^{K-1} \sum_{n=Q_m} E\left(\hat{\beta}_{mn} - \beta_{mn}\right)^2 \\
&= \sum_{n=0}^{2^M-1} \text{var}\, \hat{\alpha}_{Mn} + \sum_{m=M}^{K-1} \sum_{n\in Q_m} \text{var}\, \hat{\beta}_{mn}.
\end{aligned}
$$

Since there are 2^M empirical scaling function coefficients, $\hat{\alpha}_{Mn}$, and $q_M\,(K-M)$ empirical wavelet coefficients $\hat{\beta}_{mn}$, we get from Lemmas A.5 and A.7 that

$$\text{IV}\, \hat{g}_{MK} = \mathcal{O}\left(2^M\right) + \mathcal{O}\left(q_M\,(K-M)\right), \tag{A.13}$$

and hence, that for sufficiently large M and K and q_M set like in Theorem 5.6, for the variance of the nonlinear estimate $\hat{g}_{MK}(x)$, the following bound holds:

$$\text{IV}\, \hat{g}_{MK} = \mathcal{O}\left(\frac{2^M}{N}\right),$$

that is, the variance asymptotically does not depend on the *adaptive part* but only on the scale M of the *linear part* of the estimate (viz., on the number 2^M of scaling function empirical coefficients). This fact completes the proof of Theorem 5.5 since the approximation error vanishes as $M \to \infty$ and the error of the denominator behaves analogously.

Convergence Rates: Lemma 4.6 gives us the bound for the bias (approximation) error

$$\text{ISB}\, \hat{g}_{MK} = \text{ISE}\, g_{MK} = \mathcal{O}\left(2^{-2M}\right),$$

for the estimate with the scale $K = 2M$ (i.e., for $\nu = 2$; the lemma reveals also that the larger factors ν do not further reduce the error order). Combining both variance and approximation error bounds, we get

$$\text{MISE}\, \hat{g}_{MK} \leq \mathcal{O}\left(2^{-2M} + \frac{2^M}{N}\right),$$

which for the scale selection rule in (5.20) yields the bound for the MISE error of the estimate $\hat{g}_{MK}(x)$

$$\text{MISE}\,\hat{g}_{MK} = \mathcal{O}\left(N^{-2/3}\right).$$

Since we can easily derive the analogous bound for the MISE error of the estimate $\hat{f}_{MK}$,

$$\text{MISE}\,\hat{f}_{MK} = \mathcal{O}\left(N^{-2/3}\right),$$

then, by application of the inequality (A.5), we get *the best-case* convergence rate for the estimate $\hat{\mu}_{MK}(x)$

$$\text{MISE}\,\hat{\mu}_{MK} = \mathcal{O}\left(N^{-2/3}\right).$$

Consider now *the worst case*. The approximation error of the numerator is now equal to

$$\text{ISB}\,\hat{g}_{MK} = \text{ISE}\,g_{MK} = \mathcal{O}\left(2^{-M}\right),$$

and, for the scale selection rule as in (5.20), we obtain

$$\text{ISB}\,\hat{g}_{MK} = \mathcal{O}\left(N^{-1/3}\right).$$

The variance error remains the same (of order $N^{-2/3}$). The resulting MISE error of the NQOS estimate (since the same arguments are valid for the denominator error) converges with the slower rate of the approximation error

$$\text{MISE}\,\hat{\mu}_{MK} = \mathcal{O}\left(N^{-1/3} + N^{-2/3}\right) = \mathcal{O}\left(N^{-1/3}\right).$$

A.2 OS Algorithms

In this chapter, the proofs of the global (in the MISE error sense) convergence of the basic (linear) OS algorithm are presented in detail. The proofs of the remaining theorems (dealing with both OS and NOS algorithms) can be completed by applying the same reasoning as for the QOS-type algorithms (see the previous Appendix A.1).

A.2.1 Convergence

The mean integrated squared error (MISE) of the order statistics algorithm has a standard form

$$\text{MISE}\,\hat{\mu}_K = E \int_0^1 [\hat{\mu}_K(x) - \mu(x)]^2 \, dx$$

and can be expressed in terms of the coefficients errors

$$\text{MISE}\,\hat{\mu}_K = E \int_0^1 \left[\sum_{n=0}^{2^K-1} (\hat{\alpha}_{Kn} - \alpha_{Kn})\,\varphi_{Kn}\,(x) \right.$$

$$\left. + \sum_{m=K}^{\infty} \sum_{n=0}^{2^m-1} \beta_{mn}\psi_{mn}\,(x) \right]^2 \mathrm{d}x$$

$$\leq 2E \int_0^1 \left[\sum_{n=0}^{2^K-1} (\hat{\alpha}_{Kn} - \alpha_{Kn})\,\varphi_{Kn}\,(x) \right]^2 \mathrm{d}x$$

$$+ 2E \int_0^1 \left[\sum_{m=K}^{\infty} \sum_{n=0}^{2^m-1} \beta_{mn}\psi_{mn}\,(x) \right]^2 \mathrm{d}x.$$

Applying the orthogonality arguments (cf. (4.6) and (4.20)) simplifies the above bound to the following one

$$\text{MISE}\,\hat{\mu}_K \leq 2 \sum_{n=0}^{2^K-1} E\,(\hat{\alpha}_{Kn} - \alpha_{Kn})^2 + 2 \sum_{m=K}^{\infty} \sum_{n=0}^{2^m-1} \beta_{mn}^2, \qquad (A.14)$$

where the latter term is just the integrated approximation error and consists of all expansion terms not included in the estimate. From Sect. 4.1.11, we know that the approximation error vanishes with growing K; cf. (4.33):

$$\sum_{m=K}^{\infty} \sum_{n=0}^{2^m-1} \beta_{mn}^2 \to 0 \text{ as } K \to \infty.$$

Now, we need to examine the expectation terms. We have

$$E\,(\hat{\alpha}_{Kn} - \alpha_{Kn})^2 = E \left\{ \sum_{k=1}^{N} \int_{x_{k-1}}^{x_k} [y_k - \mu\,(x)]\,\varphi_{Kn}\,(x)\,\mathrm{d}x \right\}^2$$

$$\leq 3E \left\{ \sum_{k=1}^{N} \int_{x_{k-1}}^{x_k} [\mu\,(x_k) - \mu\,(x)]\,\varphi_{Kn}\,(x)\,\mathrm{d}x \right\}^2$$

$$+ 3E \left\{ \sum_{k=1}^{N} \xi_k \int_{x_{k-1}}^{x_k} \varphi_{Kn}\,(x)\,\mathrm{d}x \right\}^2$$

$$+ 3E \left\{ \sum_{k=1}^{N} z_k \int_{x_{k-1}}^{x_k} \varphi_{Kn}\,(x)\,\mathrm{d}x \right\}^2$$

$$= 3V_1 + 3V_2 + 3V_3.$$

To find the bounds for the latter two 'variance' terms, we need the following lemma.

Lemma A.8. *Let all the Assumption 1–4 be in force and let θ_k be a zero-mean, second-order stationary process with a variance σ_θ^2. Then the following bound holds*

$$
E\left\{\sum_{k=1}^{N}\theta_k\int_{u_{k-1}}^{u_k}\varphi_{Kn}(u)\,du\right\}^2 = \mathcal{O}\left(N^{-1}\right).
$$

Proof. Because of compactness of the wavelets support, we have

$$
E\left\{\sum_{k=1}^{N}\theta_k\int_{u_{k-1}}^{u_k}\varphi_{Kn}(u)\,du\right\}^2 = E\left\{\sum_{k=1}^{\varkappa}\theta_k\int_{u_{k-1}}^{u_k}\varphi_{Kn}(u)\,du\right\}^2,
$$

where $\varkappa = 1,\ldots,N$ is a r.v., of a *binomial distribution*. Using now a basic property of conditional expectation, we get

$$
E\left\{\sum_{k=1}^{\varkappa}\theta_k\int_{u_{k-1}}^{u_k}\varphi_{Kn}(u)\,du\right\}^2
$$

$$
= E\left\{E\sum_{i=1}^{\varkappa}\sum_{j=1}^{\varkappa}\theta_i\theta_j\int_{u_{i-1}}^{u_i}\varphi_{Kn}(u)\,du\int_{u_{j-1}}^{u_j}\varphi_{Kn}(u)\,du\,\bigg|\,\varkappa\right\},
$$

where

$$
E\left\{\sum_{i=1}^{\varkappa}\sum_{j=1}^{\varkappa}\theta_i\theta_j\int_{u_{i-1}}^{u_i}\varphi_{Kn}(u)\,du\int_{u_{j-1}}^{u_j}\varphi_{Kn}(u)\,du\,\bigg|\,\varkappa\right\}
$$

$$
= \sum_{i=1}^{\varkappa}\sum_{j=1}^{\varkappa}E\left\{\theta_i\theta_j\int_{u_{i-1}}^{u_i}\varphi_{Kn}(u)\,du\int_{u_{j-1}}^{u_j}\varphi_{Kn}(u)\,du\,\bigg|\,\varkappa\right\}.
$$

Applying the property again, we get

$$
E\left\{\theta_i\theta_j\int_{u_{i-1}}^{u_i}\varphi_{Kn}(u)\,du\int_{u_{j-1}}^{u_j}\varphi_{Kn}(u)\,du\,\bigg|\,\varkappa\right\}
$$

$$
= E\left\{\theta_i\theta_j E\left\{\int_{u_{i-1}}^{u_i}\varphi_{Kn}(u)\,du\int_{u_{j-1}}^{u_j}\varphi_{Kn}(u)\,du\,\bigg|\,\theta_i\theta_j,\varkappa\right\}\right\}.
$$

Recalling that (cf. Lemma A.3)

$$
E \left\{ \int_{u_{k-1}}^{u_k} du \right\}^2 = E \left(u_k - u_{k-1} \right)^2 = \mathcal{O}\left(N^{-2} \right),
$$

$$
E \left\{ \int_{u_{i-1}}^{u_i} du \int_{u_{j-1}}^{u_j} du \right\} = E \left(u_i - u_{i-1} \right) \left(u_j - u_{j-1} \right) = \mathcal{O}\left(N^{-2} \right)
$$

we obtain

$$
E \left\{ \int_{u_{i-1}}^{u_i} \varphi_{Kn}\left(u \right) du \int_{u_{j-1}}^{u_j} \varphi_{Kn}\left(u \right) du \,\middle|\, \theta_i \theta_j, \varkappa \right\} \le c_u 2^K N^{-2},
$$

and (since $\left| E \theta_i \theta_j \right| \le \sigma_\theta^2$ for all i, j)

$$
\sum_{i=1}^{\varkappa} \sum_{j=1}^{\varkappa} E \left\{ E \left| \theta_i \theta_j \right| E \left\{ \int_{u_{i-1}}^{u_i} \varphi_{Kn}\left(u \right) du \int_{u_{j-1}}^{u_j} \varphi_{Kn}\left(u \right) du \,\middle|\, \theta_i \theta_j, \varkappa \right\} \,\middle|\, \varkappa \right\}
$$

$$
\le c 2^K \varkappa^2 \cdot \sigma_\theta^2 N^{-2}.
$$

For $\varkappa$, we have

$$
E \left\{ \varkappa^2 \right\} \le \operatorname{var} \varkappa \le p \left(1 - p \right) N \le pN \le c 2^{-K} N,
$$

Finally,

$$
E \left\{ \sum_{k=1}^{\varkappa} \theta_k \int_{u_{k-1}}^{u_k} \varphi_{Kn}\left(u \right) du \right\}^2 \le cE \left\{ 2^K \varkappa^2 \cdot \sigma_\theta^2 N^{-2} \right\}
$$

$$
\le c 2^K 2^{-K} N \cdot \sigma_\theta^2 N^{-2} = \mathcal{O}\left(N^{-1} \right).
$$

∎

Taking $\theta_k = \xi_k$ and $\theta_k = z_k$, respectively, we obtain that

$$
V_2 = \mathcal{O}\left(N^{-1} \right) \text{ and } V_3 = \mathcal{O}\left(N^{-1} \right).
$$

Consider now the remaining 'bias' term V_1, where

$$
V_1 \le 2E \left\{ \sum_{k=1}^{N} \int_{x_{k-1}}^{x_k} \left[\mu\left(x_k \right) - \mu\left(x \right) \right] \varphi_{Kn}\left(x \right) dx \right\}^2
$$

$$
+ 2E \left\{ \int_{x_N}^{1} \left[\mu\left(x_k \right) - \mu\left(x \right) \right] \varphi_{Kn}\left(x \right) dx \right\}^2
$$

$$
= 2V_{11} + 2V_{12}.
$$

Splitting the former error V_{11} into the Lipschitz continuous and jump regions, we get

$$V_{11} \leq 2E \left\{ \sum_{k=1}^{N} \int_{x_{k-1}}^{x_k} [\mu(x_k) - \mu(x)] \varphi_{Kn}(x) \, dx \right\}^2$$

$$+ 2E \left\{ \sum_{k=1}^{q} \int_{x_{k-1}}^{x_k} [\mu(x_k) - \mu(x)] \varphi_{Kn}(x) \, dx \right\}^2$$

$$= 2V_{111} + 2V_{112},$$

In continuous regions, it holds that

$$\int_{x_{k-1}}^{x_k} |\mu(x_k) - \mu(x)| \varphi_{Kn}(x) \, dx \leq 2^{\frac{K}{2}} (x_k - x_{k-1})^2$$

Moreover, using the Cauchy–Schwarz inequality, we derive that (cf. [55, Lemma 1])

$$\left\{ \sum_{k=1}^{N} (x_k - x_{k-1})^2 \right\}^2 = \left\{ \sum_{k=1}^{N} (x_k - x_{k-1})^{\frac{1}{2}} (x_k - x_{k-1})^{\frac{3}{2}} \right\}^2$$

$$\leq \sum_{k=1}^{N} (x_k - x_{k-1}) \sum_{k=1}^{N} (x_k - x_{k-1})^3.$$

Hence, we have

$$V_{111} \leq c 2^K E \sum_{k=1}^{N} (x_k - x_{k-1})^3.$$

Exploiting again the compactness of the support of $\varphi_{Kn}(x)$ and the conditional expectation property, we get

$$V_{111} \leq c 2^K E \left\{ \sum_{k=1}^{\varkappa} E \left\{ (x_k - x_{k-1})^{2\nu+1} \middle| \varkappa \right\} \right\},$$

where (cf. again Lemma A.3)

$$E \left\{ (x_j - x_{j-1})^3 \middle| \varkappa \right\} \leq c N^{-3}.$$

Eventually, we get

$$V_{111} \leq c 2^K E \{\varkappa\} N^{-(2\nu+1)} \leq c 2^K \left(2^{-K} N \right) N^{-3} = \mathcal{O} \left(N^{-2} \right).$$

In the jump regions, we have instead

$$\int_{x_{k-1}}^{x_k} |\mu(x_k) - \mu(x)| \, \varphi_{Kn}(x) \, dx \le c_\mu 2^{\frac{K}{2}} (x_k - x_{k-1}).$$

Recalling that

$$E\left\{(x_i - x_{i-1})(x_j - x_{j-1})\right\} \le c_2 N^{-2},$$

for some $c_2 > 0$, and denoting now by $\varkappa = 1, \ldots, q$ a r.v. being a number of jumps inside the support of $\varphi_{Kn}(x)$, we get that

$$V_{112} \le 2^K E \left\{ \sum_{i=1}^{\varkappa} \sum_{j=1}^{\varkappa} E\left\{(x_i - x_{i-1})(x_j - x_{j-1}) \,|\, \varkappa\right\} \right\}$$

$$\le c 2^K E\left\{\varkappa^2\right\} \cdot N^{-2} \le c 2^K \cdot q 2^{-K} \cdot N^{-2} = cq N^{-2}.$$

Consider the latter term, V_{12}, related to the spacing between the last (ordered) measurement x_N and the right boundary point $x = 1$. Applying the Cauchy–Schwarz inequality, we have

$$V_{12} \le E \left\{ \left[\int_{x_N}^1 [\mu(x_k) - \mu(x)] \, \varphi_{Kn}(x) \, dx \right]^2 \right\}$$

$$\le E \left\{ \int_{x_N}^1 [\mu(x_k)(x)]^2 \, dx \int_{x_N}^1 \varphi_{Kn}^2(x) \, dx \right\}$$

$$\le E \int_{x_N}^1 [\mu(x_k) - \mu(x)]^2 \, dx,$$

which yields $\mathcal{O}(N^{-1})$ or $\mathcal{O}(N^{-3})$ if $\mu(x)$ has (or does not have) jumps in the interval $(x_N, 1)$, respectively. Eventually, we get that $V_1 = \mathcal{O}(N^{-1})$, and we conclude that for any piecewise-Lipschitz nonlinearity $\mu(x)$, it holds that

$$E(\hat{\alpha}_{Kn} - \alpha_{Kn})^2 = \mathcal{O}(N^{-1}) \quad \text{and} \quad \sum_{n=0}^{2^K - 1} E(\hat{\alpha}_{Kn} - \alpha_{Kn})^2 = \mathcal{O}\left(\frac{2^K}{N}\right). \quad (A.15)$$

In consequence, if $K \to \infty$ as $N \to \infty$ so slow that $2^K/N \to 0$, then

$$\text{MISE} \, \hat{\mu}_K \to 0 \text{ as } N \to \infty,$$

and the proof of the linear OS algorithm convergence Theorem 5.7 is completed.

A.3 ED Algorithms

We need some preliminary results concerning the smoothness of the composite function $\mu_F(x)$. Denote $x = F(u)$ and $u = F^{-1}(x)$ and observe that:

- $F(u)$ has a derivative (a density function, $f(u)$), which, by Assumption 1, is strictly positive, i.e., $F'(u) = f(u) \geq c$, some $c > 0$.
- The inverse $F^{-1}(x)$ has a derivative

$$f^{-1}(x) = \left[F^{-1}\right]'(x) = \frac{1}{f\left(F^{-1}(x)\right)}$$

  which is bounded (again, by virtue of the assumption that the density, $f(u)$, is strictly positive).
- $F^{-1}(x)$ is Lipschitz continuous (since its derivative is bounded).

Hence, we get the lemma.

Lemma A.9. *If the nonlinearity $\mu(u)$ is piecewise-Lipschitz, then so is $\mu_F(x)$, that is,*

$$|\mu_F(w) - \mu_F(v)| \leq c\,|w - v|, \text{ for some constant } c > 0.$$

Proof. We have

$$\mu(u) = \mu\left(F^{-1}(x)\right) = \mu_F\left(F\left(F^{-1}(x)\right)\right) = \mu_F(F(u)) = \mu_F(x)$$

and hence,

$$\begin{aligned}|\mu_F(w) - \mu_F(v)| &= \left|\mu\left(F^{-1}(w)\right) - \mu\left(F^{-1}(v)\right)\right| \\ &\leq c\left|F^{-1}(w) - F^{-1}(v)\right| \\ &\leq c\,|w - v|,\end{aligned}$$

for each w, v in all Lipschitz regions of $\mu(u)$. ∎

Remark A.1. Note that the inverse of the cumulative distribution function (or its empirical version) is not *explicitly* used in the algorithm.

We begin with the analysis of the MISE error of the estimate $\hat{\mu}_K(F(u)) = \hat{\mu}_F(x)$, where $F(u)$ is a distribution function of the input u_k. We have here

$$\text{MISE}\,\hat{\mu}_F = E \int_0^1 \left[\hat{\mu}_F(x) - \mu_F(x)\right]^2 dx. \tag{A.16}$$

Observe that $x = F(u)$ (and hence $dx = f(u)\,du$). Moreover, the integration limits are $0 = F(u_{\min})$, and $1 = F(u_{\max})$, and hence,

$$\text{MISE } \hat{\mu}_F = E \int_{u_{\min}}^{u_{\max}} [\hat{\mu}_F (F (u)) - \mu (u)]^2 f (u) \, du$$

$$= E \int_0^1 [\hat{\mu}_F (F (u)) - \mu (u)]^2 f (u) \, du,$$

since by Assumption 1, we have also that $u_{\min} = 0$ and $u_{\max} = 1$. Furthermore, we have $0 < c \le f (u) \le C$, for some constants c, C, which yields that

$$C \text{ MISE } \hat{\mu}_F \ge E \int_0^1 [\hat{\mu}_F (F (u)) - \mu (u)]^2 \, du \ge c \text{ MISE } \hat{\mu}_F,$$

and to show the asymptotic properties of the ED algorithm it suffices to examine the error in (A.16).

We also need to inspect the fact that we use the empirical distribution instead of the genuine one, *that is*, that in fact $\hat{\mu} (u) = \hat{\mu}_F (F_N (u)) = \hat{\mu}_F (\lfloor Nx \rfloor / N)$, and hence, we should rather consider the following MISE error allowing for the *binning error* (cf. e.g. [66, 117, Chap. 12]):

$$\text{MISE } \hat{\mu}_F = E \int_0^1 \left[\hat{\mu}_F \left(\tfrac{\lfloor Nx \rfloor}{N} \right) - \mu_F (x) \right]^2 \, dx.$$

Observe, however, that we assumed that the number of measurements, N, is a *dyadic integer* and $\lfloor Nx \rfloor / N$ is simply the representation of x *truncated* after $\log_2 N$ bits. Thus, since $K \le \log_2 N$, then if $x \in \text{supp} \, \varphi_{Kn}$, then $\lfloor Nx \rfloor / N \in \text{supp} \, \varphi_{Kn}$ for all x. In consequence, we have that

$$\hat{\mu}_F \left(\tfrac{\lfloor Nx \rfloor}{N} \right) - \hat{\mu}_F (x) = \sum_{n=0}^{2^K - 1} \hat{\alpha}_{Kn} \left[\varphi_{Kn} \left(\tfrac{\lfloor Nx \rfloor}{N} \right) - \varphi_{Kn} (x) \right] = 0,$$

for all $x \in [0, 1]$, that is, the binning error is zero and the MISE error formula in (A.16) exactly describes our ED estimate.

A.3.1 Convergence

The MISE error of the ED estimate can be decomposed using the following terms (cf. (A.14)):

$$\text{MISE } \hat{\mu}_K \le 2 \sum_{n=0}^{2^K - 1} E \, (\hat{\alpha}_{Kn} - \alpha_{Kn})^2 + 2 \sum_{m=K}^{\infty} \sum_{n=0}^{2^m - 1} \beta_{mn}^2.$$

One can easily ascertain that the latter, 'approximation' term, vanishes as K tends to infinity (cf. Sect. 4.1.11) since (from Lemma A.9) the nonlinearity $\mu_F(u)$ derives its boundedness and smoothness properties from $\mu(u)$. In what follows, we examine the 'variance' part in detail. Recall that we have the following equivalence, cf. (5.34),

$$\hat{\alpha}_{Kn} = \sum_{k=1}^{N} y_k \int_{\frac{k-1}{N}}^{\frac{k}{N}} \varphi_{Kn}(\upsilon)\, d\upsilon = \sum_{k=1}^{N} y_k \int_{F_N(u_{k-1})}^{F_N(u_k)} \varphi_{Kn}(\upsilon)\, d\upsilon.$$

Moreover, it holds that

$$\alpha_{kn} = \int_0^1 \mu_F(\upsilon)\, \varphi_{Kn}(\upsilon)\, d\upsilon = \int_{F(0)}^{F(1)} \mu_F(\upsilon)\, \varphi_{Kn}(\upsilon)\, d\upsilon$$

$$= \sum_{k=1}^{N} \int_{F(u_{k-1})}^{F(u_k)} \mu_F(\upsilon)\, \varphi_{Kn}(\upsilon)\, d\upsilon,$$

and thus for the input-output measurements, we have

$$y_k = \mu(u_k) + \xi_k + z_k = \mu_N(u_k) + \xi_k + z_k$$

$$= \mu \circ F_N(u_k) + \xi_k + z_k, \tag{A.17}$$

where $\mu_N(u) = \mu \circ F_N(u)$ is the composite function of the nonlinearity and the empirical distribution. Observe that in all measurement points u_k, we have that $\mu_N(u_k) = \mu(u_k)$, that is, $\mu_N(u_k)$ is equal to the genuine nonlinearity in these points. Denote by $\mu_N(u) = \mu \circ F_N(u)$ the composite functions of the nonlinearity and of the empirical distribution $F_N(u)$. The following decomposition of the empirical coefficients in the ED algorithm holds true

$$\hat{\alpha}_{Kn} - \alpha_{kn}$$

$$= \underbrace{\sum_{k=1}^{N} y_k \int_{F_N(u_{k-1})}^{F_N(u_k)} \varphi_{Kn}(\upsilon)\, d\upsilon - \sum_{k=1}^{N} \mu_N(u_k) \int_{F_N(u_{k-1})}^{F_N(u_k)} \varphi_{Kn}(\upsilon)\, d\upsilon}_{\text{variance error}}$$

$$+ \underbrace{\sum_{k=1}^{N} \mu_N(u_k) \int_{F_N(u_{k-1})}^{F_N(u_k)} \varphi_{Kn}(\upsilon)\, d\upsilon - \sum_{k=1}^{N} \mu_F(u_k) \int_{F_N(u_{k-1})}^{F_N(u_k)} \varphi_{Kn}(\upsilon)\, d\upsilon}_{\text{bias error}}$$

$$+\sum_{k=1}^{N} \mu_F\left(u_k\right) \int_{F_N\left(u_{k-1}\right)}^{F_N\left(u_k\right)} \varphi_{Kn}\left(\upsilon\right) d\upsilon - \sum_{k=1}^{N} \int_{F_N\left(u_{k-1}\right)}^{F_N\left(u_k\right)} \mu_F\left(\upsilon\right) \varphi_{Kn}\left(\upsilon\right) d\upsilon$$

$$\underbrace{}_{\text{approximation error}}$$

$$+\sum_{k=1}^{N} \int_{F_N\left(u_{k-1}\right)}^{F_N\left(u_k\right)} \mu_F\left(\upsilon\right) \varphi_{Kn}\left(\upsilon\right) d\upsilon - \sum_{k=1}^{N} \int_{F\left(u_{k-1}\right)}^{F\left(u_k\right)} \mu_F\left(\upsilon\right) \varphi_{Kn}\left(\upsilon\right) d\upsilon.$$

$$\underbrace{}_{\text{empirical distribution error}}$$

And hence, using (A.17), we get

$$\hat{\alpha}_{Kn} - \alpha_{kn} = \sum_{k=1}^{2^{-K}N} \xi_k \int_{\frac{k-1}{N}}^{\frac{k}{N}} \varphi_{Kn}\left(\upsilon\right) d\upsilon$$

$$+ \sum_{k=1}^{2^{-K}N} z_k \int_{\frac{k-1}{N}}^{\frac{k}{N}} \varphi_{Kn}\left(\upsilon\right) d\upsilon$$

$$+ \sum_{k=1}^{2^{-K}N} \left[\mu_N\left(u_k\right) - \mu_F\left(u_k\right)\right] \int_{\frac{k-1}{N}}^{\frac{k}{N}} \varphi_{Kn}\left(\upsilon\right) d\upsilon$$

$$+ \sum_{k=1}^{2^{-K}N} \left[\mu_F\left(u_k\right) - \mu_F\left(\tfrac{k}{N}\right)\right] \int_{\frac{k-1}{N}}^{\frac{k}{N}} \varphi_{Kn}\left(\upsilon\right) d\upsilon$$

$$+ \sum_{k=1}^{2^{-K}N} \int_{\frac{k-1}{N}}^{\frac{k}{N}} \left[\mu_F\left(\tfrac{k}{N}\right) - \mu_F\left(\upsilon\right)\right] \varphi_{Kn}\left(\upsilon\right) d\upsilon$$

$$+ \sum_{k=1}^{2^{-K}N} \int_{F\left(u_k\right)}^{\frac{k}{N}} \mu_F\left(\upsilon\right) \varphi_{Kn}\left(\upsilon\right) d\upsilon$$

$$+ \sum_{k=1}^{2^{-K}N} \int_{F\left(u_{k-1}\right)}^{\frac{k-1}{N}} \mu_F\left(\upsilon\right) \varphi_{Kn}\left(\upsilon\right) d\upsilon$$

$$= V_1 + \cdots + V_7, \text{ say.}$$

For each of the V-terms, we have respectively that (see Lemmas A.2 and A.3; cf. the proof of Lemma A.8):

$$EV_1^2 = 2^K E \left[\sum_{k=1}^{2^{-K}N} \xi_k \int_{\frac{k-1}{N}}^{\frac{k}{N}} dv \right]^2 \leq cN^{-1},$$

$$EV_2^2 = 2^K E \left[\sum_{k=1}^{2^{-K}N} z_k \int_{\frac{k-1}{N}}^{\frac{k}{N}} dv \right]^2 \leq cN^{-1}.$$

Moreover, recalling that $\mu(u)$ and $F(u)$ are both Lipschitz and that $\hat{F}_N(u)$ is an empirical distribution of $F(u)$ (i.e., its unbiased estimate), we derive that for each $u \in [0, 1]$, it holds that $E\left[\mu(F(u)) - \mu(\hat{F}_N(u))\right]^2 \leq cN^{-1}$ and subsequently that

$$EV_3^2 = 2^K E \left[\sum_{k=1}^{2^{-K}N} [\mu_N(u_k) - \mu_F(u_k)] \int_{\frac{k-1}{N}}^{\frac{k}{N}} dv \right]^2 \leq cN^{-1}.$$

Furthermore,

$$EV_4^2 = 2^K E \left[\sum_{k=1}^{2^{-K}N} [\mu_F(u_k) - \mu_F(\tfrac{k}{N})] \int_{\frac{k-1}{N}}^{\frac{k}{N}} dv \right]^2 \leq cN^{-1},$$

$$EV_5^2 = 2^K E \left[\sum_{k=1}^{2^{-K}N} \int_{\frac{k-1}{N}}^{\frac{k}{N}} [\mu_F(\tfrac{k}{N}) - \mu_F(v)] dv \right]^2 \leq cN^{-1}.$$

Using the same arguments, we infer that for the 'empirical distribution' terms, we have that

$$EV_6^2 = 2^K E \left[\sum_{k=1}^{2^{-K}N} \int_{F(u_k)}^{\frac{k}{N}} \mu_F(v) dv \right]^2 \leq cN^{-1},$$

$$EV_7^2 = 2^K E \left[\sum_{k=1}^{2^{-K}N} \int_{F(u_{k-1})}^{\frac{k-1}{N}} \mu_F(v) dv \right]^2 \leq cN^{-1},$$

and hence, for each empirical coefficient $\hat{\alpha}_{Kn}$ in the ED algorithm, we have that

$$E(\hat{\alpha}_{Kn} - \alpha_{Kn})^2 = \mathcal{O}(N^{-1}).$$

which ends the proof of the convergence of the ED algorithm, since one can easily determine that (cf. A.15)

$$E\left(\hat{\alpha}_{Kn} - \alpha_{Kn}\right)^2 = \mathcal{O}\left(N^{-1}\right) \text{ and } \sum_{n=0}^{2^K-1} E\left(\hat{\alpha}_{Kn} - \alpha_{Kn}\right)^2 = \mathcal{O}\left(\frac{2^K}{N}\right).$$

The remaining convergence and convergence rate theorems for both linear and nonlinear versions of the empirical distribution algorithm can be proven using the reasoning applied to the corresponding results for the QOS-type algorithms; see Appendix A.1. In particular, recall that by virtue of Lemma A.9, if $\mu\left(u\right)$ is (piecewise-)Lipschitz, then so is the nonlinearity $\mu_F\left(u\right) = \mu \circ F\left(u\right)$.

A.4   EOS Algorithms

The analysis of the EOS algorithm convergence properties differs from those performed for previous algorithms in that the unbalanced wavelet basis, being generated by the measurements set, is not known a priori and in fact is random; cf. Sect. 4.2.1 and see [41]. The proof scenario remains nevertheless similar thanks to the observations that—in spite of this randomness—the compact supports of the unbalanced Haar basis functions vanish as in the classic Haar case (only slower by a logarithmic factor; cf. 4.51).

A.4.1   Convergence

We start with a decomposition of the MISE error:

$$\text{MISE } \hat{\mu}_K = E \int_0^1 \left[\hat{\mu}_k\left(x\right) - \mu\left(x\right)\right]^2 dx$$

$$= 2E \int \left[\sum_{n=0}^{2^K-1} \left(\hat{\alpha}_{Kn} - a_{Kn}\right) \varphi_{Kn}\left(x\right)\right]^2 dx$$

$$+ 2E \int \left[\sum_{n=0}^{2^K-1} a_{Kn} \varphi_{Kn}\left(x\right) - \mu\left(x\right)\right]^2 dx$$

For the former, we know from Sect. 4.2.4 that it vanishes as $K \to \infty$ (along with $N \to \infty$). Applying the orthonormality property of the unbalanced Haar basis, we get for the first term that

$$E \int \left[\sum_{n=0}^{2^K-1} \left(\hat{\alpha}_{Kn} - a_{Kn}\right) \varphi_{kn}\left(x\right)\right]^2 dx = E \sum_{n=0}^{2^K-1} \left(\hat{\alpha}_{Kn} - a_{Kn}\right)^2.$$

And we can now focus on the expectation term

$$
E \left(\hat{\alpha}_{Kn} - a_{Kn} \right)^2 = E \left[\sum_{k=1}^{N} \int_{x_{k-1}}^{x_k} \left[y_k - \mu \left(x \right) \right] \varphi_{Kn} \left(x \right) \mathrm{d}x \right]^2
$$

$$
\leq 3E \left[\sum_{k=1}^{N} \int_{x_{k-1}}^{x_k} \left[\mu \left(x_k \right) - \mu \left(x \right) \right] \varphi_{Kn} \left(x \right) \mathrm{d}x \right]^2
$$

$$
+ 3E \left[\sum_{k=1}^{N} \int_{x_{k-1}}^{x_k} \xi_k \varphi_{Kn} \left(x \right) \mathrm{d}x \right]^2
$$

$$
+ 3E \left[\sum_{k=1}^{N} \int_{x_{k-1}}^{x_k} z_k \varphi_{Kn} \left(x \right) \mathrm{d}x \right]^2
$$

$$
= 3 \left(V_1 + V_2 + V_3 \right)
$$

where for the former term we have

$$
V_1 = \sum_{i=1}^{2^{-K}N} \sum_{j=1}^{2^{-K}N} E \left\{ \int_{x_{i-1}}^{x_i} \left[\mu \left(x_i \right) - \mu \left(x \right) \right] \varphi_{Kn} \left(x \right) \mathrm{d}x \right.
$$

$$
\left. \times \int_{x_{j-1}}^{x_j} \left[\mu \left(x_j \right) - \mu \left(x \right) \right] \varphi_{Kn} \left(x \right) \mathrm{d}x \right\}
$$

with

$$
\int_{x_{i-1}}^{x_i} \left| \mu \left(x_i \right) - \mu \left(x \right) \right| \varphi_{Kn} \left(x \right) \mathrm{d}x \leq c \sqrt{I_k^{-1}} I_k^2 = c I_k^{\frac{3}{2}},
$$

for Lipschitz regions, and

$$
\int_{x_{i-1}}^{x_i} \left| \mu \left(x_i \right) - \mu \left(x \right) \right| \varphi_{Kn} \left(x \right) \mathrm{d}x \leq c \sqrt{I_k^{-1}} I_k = c I_k^{\frac{1}{2}}
$$

in the intervals with jumps. Since we know that with probability one (cf. (4.51)) $I_k = \mathcal{O} \left(N^{-1} \log N \right)$, then

$$
V_1 \leq c \sum_{i=1}^{2^{-K}N} \sum_{j=1}^{2^{-K}N} I_k^3 + c \sum_{i=1}^{q} \sum_{j=1}^{q} I_k = \mathcal{O} \left(2^{-2K} \frac{\log^3 N}{N} + \frac{\log N}{N} \right).
$$

Since $K = 3^{-1} \log_2 N$, then

$$V_1 = \mathcal{O}\left(\frac{\log^2 N}{N^{2/3}} \frac{\log N}{N} + \frac{\log N}{N}\right) = \mathcal{O}\left(\frac{\log N}{N}\right).$$

Examining V_2, we have due to Lemma A.2 that

$$V_2 \leq \sum_{i=1}^{2^{-K}N} \sum_{j=1}^{2^{-K}N} E\left\{|\xi_i \xi_j| \int_{x_{i-1}}^{x_i} \varphi_{Kn}(x)\,dx \int_{x_{j-1}}^{x_j} \varphi_{Kn}(x)\,dx\right\}$$

$$\leq c\frac{1}{N} E \int_{x_{i-1}}^{x_i} \varphi_{Kn}(x)\,dx \int_{x_{j-1}}^{x_j} \varphi_{Kn}(x)\,dx.$$

Applying again (4.51), we get w.p.1

$$E \int_{x_{i-1}}^{x_i} \varphi_{Kn}(x)\,dx \int_{x_{j-1}}^{x_j} \varphi_{Kn}(x)\,dx \leq c2^K \frac{\log^2 N}{N^2}.$$

And hence,

$$V_2 \leq c \sum_{i=1}^{2^{-K}N} \sum_{j=1}^{2^{-K}N} \frac{1}{N} 2^K \frac{\log^2 N}{N^2} = \mathcal{O}\left(\frac{\log^2 N}{N}\right).$$

A similar bound holds for V_3, and thus, we have established that for any K and any n, we have

$$E\left(\hat{\alpha}_{Kn} - a_{Kn}\right)^2 = \mathcal{O}\left(\frac{\log N}{N}\right) + \mathcal{O}\left(\frac{\log^2 N}{N}\right) = \mathcal{O}\left(\frac{\log^2 N}{N}\right),$$

and hence

$$\sum_{n=0}^{2^K-1} E\left(\hat{\alpha}_{Kn} - a_{Kn}\right)^2 = \mathcal{O}\left(2^K \frac{\log^2 N}{N}\right).$$

Collecting now the above bounds and the results from Sect. 4.2.4 concerning the convergence rates of the approximation error of the unbalance approximant for Lipschitz and piecewise-Lipschitz nonlinearities, one can complete the proofs of the remaining theorems from Sect. 5.8 as for the previous algorithms.

Appendix B
Common Symbols

Notation	Name
$\varphi(x)$	Haar scaling function (*father wavelet*)
$\psi(x)$	Haar wavelet (*mother wavelet*)
$\phi(x)$	Haar reproducing kernel
$\Phi(x)$	Indefinite integral of $\varphi(x)$
$\varphi_{mn}(x), \psi_{mn}(x)$	Scaled and translated wavelet functions
V_m	Approximation space
W_m	Detail space
$\chi_{[a,b]}(x)$	Indicator function of the interval $[a, b]$
$\chi_k, \chi_k(x)$	kth sample block and its indicator function
χ_{mn}	Support of unbalanced Haar functions $\varphi_{mn}(x), \psi_{mn}(x)$
$\chi_{mn}(x)$	Indicator function of χ_{mn}
I_k	kth spacing, length of χ_k
I_{mn}	Length of χ_{mn}
$m(u)$	Nonlinear characteristic
$\{\lambda_i\}$	Linear dynamics impulse response
$\mu(u)$	Identified system nonlinearity
u_k	System inputs
y_k	System outputs
z_k	External noise
ξ_k	'System' noise
x_k	Algorithm inputs
N	Number of measurement pairs (u_k, y_k) (or (x_k, y_k))
$\hat{\mu}_K(x)$	Linear estimate

P. Śliwiński, *Nonlinear System Identification by Haar Wavelets*, Lecture Notes in Statistics 210, DOI 10.1007/978-3-642-29396-2,
© Springer-Verlag Berlin Heidelberg 2013

$\hat{\mu}_{MK}(x)$	Nonlinear estimate
α_{mn}, β_{mn}	Scaling function and wavelet expansion coefficients
$\hat{\alpha}_{mn}, \hat{\beta}_{mn}$	Estimates of α_{mn}, β_{mn} (*empirical coefficients*)
Q_m	Set of indices n of the empirical coefficients $\hat{\beta}_{mn}$
	Residing in the empirical cones of influence

References

1. Apostol, T.M.: Calculus, Volume 1. One-Variable Calculus with an Introduction to Linear Algebra, 2nd edn. Wiley, New York (1975)
2. Balestrino, A., Landi, A., Ould-Zmirli, M., Sani, L.: Automatic nonlinear auto-tuning method for Hammerstein modeling of electrical drives. IEEE Transactions on Industrial Electronics **48**(3), 645–655 (2001)
3. Bayer, R.: Symmetric binary B-Trees: Data structure and maintenance algorithms. Acta Informatica **1**(4), 290–306 (1972)
4. Bendat, J.S.: Nonlinear System Analysis and Identification. Wiley, New York (1990)
5. Billings, S.A.: Identification of non-linear systems—a survey. Proceedings of IEE **127**(6), 272–285 (1980)
6. Billings, S.A., Fakhouri, S.Y.: Theory of separable processes with application to the identification of non-linear systems. Proceedings of IEE **125**(10), 1051–1058 (1978)
7. Billings, S.A., Fakhouri, S.Y.: Non-linear system identification using the Hammerstein model. International Journal of Systems Science **10**, 567–578 (1979)
8. Blu, T., Thévenaz, P., Unser, M.: Linear interpolation revitalized. IEEE Transactions on Image Processing **13**(5), 710–719 (2004)
9. Blu, T., Unser, M.: Wavelets, fractals, and radial basis functions. IEEE Transactions on Signal Processing **50**(3), 543–553 (2002)
10. Boyd, S., Chua, L.: Fading memory and the problem of approximating nonlinear operators with Volterra series. Circuits and Systems, IEEE Transactions on **32**(11), 1150–1161 (1985)
11. Buhmann, M.D.: Radial basis functions. Acta Numerica **9**, 1–38 (2001)
12. Buhmann, M.D.: Radial Basis Functions: Theory and Implementations. Cambridge University Press, Cambridge (2003)
13. Capobianco, E.: Hammerstein system representation of financial volatility processes. The European Physical Journal B - Condensed Matter **27**(2), 201–211 (2002)
14. Chen, H.F.: Pathwise convergence of recursive identification algorithms for Hammerstein systems. IEEE Transactions on Automatic Control **49**(10), 1641–1649 (2004)
15. Chen, H.F.: Strong consistency of recursive identification for Hammerstein systems with discontinuous piecewise-linear memoryless block. IEEE Transactions on Automatic Control **50**(10), 1612–1617 (2005)
16. Cohen, A.: Numerical analysis of wavelets methods. Studies in Mathematics and Its Applications. Elsevier, Amsterdam (2003)
17. Cohen, A.: Theoretical, applied and computational aspects of nonlinear approximation. In: J. Bramble, A. Cohen, W. Dahmen, C. Canuto (eds.) Multiscale Problems and Methods in Numerical Simulations, *Lecture Notes in Mathematics*, vol. 1825/2003, pp. 1–29. Springer-Verlag, Berlin Heidelberg (2003)

18. Cohen, A., D'Ales, J.P.: Nonlinear approximation of random functions. SIAM Journal of Applied Mathematics **57**(2), 518–540 (1997)
19. Cohen, A., Daubechies, I., Vial, P.: Wavelet bases on the interval and fast algorithms. Journal of Applied and Computational Harmonic Analysis **1**(1), 54–81 (1993)
20. Cripps, S.: RF power amplifiers for wireless communications. Artech House (2006)
21. Dahlquist, G., Björk, A.: Numerical Methods. Prentice-Hall, Inc. Englewood Cliffs, New Jersey (1974)
22. Daubechies, I.: Orthonormal bases of compactly supported wavelets. Communication on Pure and Applied Mathematics **42**, 909–996 (1988)
23. Daubechies, I.: Ten Lectures on Wavelets. SIAM Edition, Philadelphia (1992)
24. Daubechies, I., Sweldens, W.: Factoring wavelet transforms into lifting steps. The Journal of Fourier Analysis and Applications **4**(3), 245–267 (1998)
25. David, H.A., Nagaraja, H.N.: Order statistics, 3rd edn. John Wiley & Sons, Inc., Hoboken, New Jersey (2003)
26. Delouille, V., Franke, L., von Sachs, R.: Nonparametric stochastic regression with design-adapted wavelets. Sankhyā, Ser. A **63**(3), 328–366 (2001)
27. Delouille, V., Simoens, J., von Sachs, R.: Smooth design-adapted wavelets for nonparametric stochastic regression. Journal of the American Statistical Association **99**(467), 643–658 (2004)
28. Dempsey, E., Westwick, D.: Identification of Hammerstein models with cubic spline nonlinearities. IEEE Transactions on Biomedical Engineering **51**(2), 237–245 (2004)
29. DeVore, R.A.: Nonlinear approximation. Acta Numerica **7**, 51–150 (1998)
30. DeVore, R.A.: Optimal computation. Proceedings of the International Congress of Mathematicians, Madrid, Spain, 2006 (2006)
31. DeVore, R.A., Lorentz, G.G.: Constructive Approximation. Springer-Verlag, Berlin Heidelberg New York (1993)
32. DeVore, R.A., Lucier, B.J.: Wavelets. Acta Numerica **1**, 1–56 (1992)
33. Donoho, D.L.: De-noising by soft thresholding. IEEE Transactions on Information Theory **41**(3), 613–627 (1995)
34. Donoho, D.L., Johnstone, I.M.: Ideal spatial adaptation via wavelet shrinkage. Biometrika **81**, 425–455 (1994)
35. Donoho, D.L., Johnstone, I.M.: Adapting to unknown smoothness via wavelet shrinkage. Journal of the American Statistical Association **90**, 1200–1224 (1995)
36. Donoho, D.L., Johnstone, I.M.: Minimax estimation via wavelets shrinkage. Annals of Statistics **26**, 879–921 (1998)
37. Ferrari, S., Maggioni, M., Borghese, N.A.: Multiscale approximation with hierarchical radial basis functions networks. IEEE Transaction one Neural Networks **15**(1) (2004)
38. Gallman, P.: An iterative method for the identification of nonlinear systems using a Uryson model. IEEE Transactions on Automatic Control **20**(6), 771–775 (1975)
39. Giannakis, G.B., Serpedin, E.: A bibliography on nonlinear system identification. Signal Processing **81**(3), 533–580 (2001)
40. Gilabert, P., Montoro, G., Bertran, E.: FPGA implementation of a real-time NARMA-based digital adaptive predistorter. IEEE Transactions on Circuits and Systems II: Express Briefs **58**(7), 402–406 (2011)
41. Girardi, M., Sweldens, W.: A new class of unbalanced Haar wavelets that form an unconditional basis for l_p on general measure spaces. The Journal of Fourier Analysis and Applications **3**(4), 457–474 (1997)
42. Giri, F., Bai, E.W. (eds.): Block-oriented nonlinear system identification. Lecture Notes in Control and Information Sciences. Springer-Verlag, Berlin Heidelberg (2010)
43. Gomes, S.M., Cortina, E.: Some results on the convergence of sampling series based on convolution integrals. SIAM Journal on Mathematical Analysis **26**(5), 1386–1402 (1995)
44. Graham, R.L., Knuth, D.E., Patashnik, O.: Concrete Mathematics. Addison-Wesley, Reading, Massachusetts (1994)

45. Gray, R.M., Neuhoff, D.L.: Quantization. IEEE Transactions on Information Theory **44**(6), 2325–2383 (1998)
46. Greblicki, W.: Nonparametric system identification by orthogonal series. Problems of Control and Information Theory **8**, 67–73 (1979)
47. Greblicki, W.: Nonparametric orthogonal series identification of Hammerstein systems. International Journal of Systems Science **20**(12), 2355–2367 (1989)
48. Greblicki, W.: Nonparametric identification of Wiener systems. IEEE Transactions on Information Theory **38**(5), 1487–1493 (1992)
49. Greblicki, W.: Nonparametric approach to Wiener system identification. IEEE Transactions on Circuits and Systems - I: Fundamental Theory and Applications **44**, 538–545 (1997)
50. Greblicki, W.: Nonlinearity recovering in Wiener system driven with correlated signal. IEEE Transactions on Automatic Control **49**(10), 1805–1810 (2004)
51. Greblicki, W.: Nonparametric input density-free estimation of the nonlinearity in Wiener systems. Information Theory, IEEE Transactions on **56**(7), 3575–3580 (2010)
52. Greblicki, W., Pawlak, M.: Identification of discrete Hammerstein system using kernel regression estimates. IEEE Transactions on Automatic Control **31**(1), 74–77 (1986)
53. Greblicki, W., Pawlak, M.: Nonparametric identification of Hammerstein systems. IEEE Transactions on Information Theory **35**, 409–418 (1989)
54. Greblicki, W., Pawlak, M.: Recursive nonparametric identification of Hammerstein systems. Journal of the Franklin Institute **326**(4), 461–481 (1989)
55. Greblicki, W., Pawlak, M.: Dynamic system identification with order statistics. IEEE Transactions on Information Theory **40**, 1474–1489 (1994)
56. Greblicki, W., Pawlak, M.: Nonparametric recovering nonlinearities in block oriented systems with the help of Laguerre polynomials. Control – Theory and Advanced Technology **10**(4), 771–791 (1994)
57. Greblicki, W., Pawlak, M.: Nonparametric System Identification. Cambridge University Press, New York (2008)
58. Györfi, L., Kohler, M., A. Krzyżak, Walk, H.: A Distribution-Free Theory of Nonparametric Regression. Springer-Verlag, New York (2002)
59. Haar, A.: Zur Theorie der Orthogonalen Funktionen-Systeme. Annals of Mathematics **69** (1910)
60. Haar, A.: On the theory of orthogonal function systems. In: C. Heil, D.F. Walnut (eds.) Fundamental papers in wavelet theory, pp. 155–188. Priceton University Press, Princeton and Oxford (2006)
61. Haber, R., Keviczky, L.: Nonlinear System Parameter Identification. Kluwer Academic Publishers, Dordrecht-Boston-London (1999)
62. Hall, P., Kerkyacharian, G., Picard, D.: Block threshold rules for curve estimation using kernel and wavelet methods. The Annals of Statistics **26**(3), 922–942 (1998)
63. Hall, P., Patil, P.: On the choice of smoothing parameter, threshold and truncation in nonparametric regression by non-linear wavelet methods. Journal of the Royal Statistical Society. Series B (Methodological) **58**(2), 361–377 (1996)
64. Hall, P., Turlach, B.: Interpolation methods for nonlinear wavelet regression with irregularly spaced design. The Annals of Statistics **25**(5), 1912–1925 (1997)
65. Härdle, W.: Applied Nonparametric Regression. Cambridge University Press, Cambridge (1990)
66. Härdle, W., Kerkyacharian, G., Picard, D., Tsybakov, A.: Wavelets, Approximation, and Statistical Applications. Springer-Verlag, New York (1998)
67. Härdle, W., Müller, M., Sperlich, S., Werwatz, A.: Nonparametric and Semiparametric Models. Springer-Verlag, Berlin Heidelberg (2004)
68. Hasiewicz, Z.: Hammerstein system identification by the Haar multiresolution approximation. International Journal of Adaptive Control and Signal Processing **13**(8), 697–717 (1999)
69. Hasiewicz, Z.: Modular neural networks for non-linearity recovering by the Haar approximation. Neural Networks **13**, 1107–1133 (2000)

70. Hasiewicz, Z.: Wavelet network for recursive function learning. In: Neural Networks and Soft Computing, Advances in Soft Computing, pp. 710–715. 6th IEEE International Conference on Neural Networks and Soft Computing, Physica-Verlag, Springer-Verlag Company, Heidelberg, Zakopane 2002 (2003)

71. Hasiewicz, Z., Pawlak, M., Śliwiński, P.: Non-parametric identification of non-linearities in block-oriented complex systems by orthogonal wavelets with compact support. IEEE Transactions on Circuits and Systems I: Regular Papers 52(1), 427–442 (2005)

72. Hasiewicz, Z., Śliwiński, P.: Identification of non-linear characteristics of a class of block-oriented non-linear systems via Daubechies wavelet-based models. International Journal of Systems Science 33(14), 1121–1144 (2002)

73. Heil, C., Walnut, D.F. (eds.): Fundamental papers in wavelet theory. Priceton University Press, Priceton and Oxford (2006)

74. Hunter, I.W., Korenberg, M.J.: The identification of non-linear biological systems: Wiener and Hammerstein cascade models. Biological Cybernetics 55(2–3), 135–144 (1986)

75. Huoa, H.B., Zhonga, Z.D., Zhua, X.J., Tua, H.Y.: Nonlinear dynamic modeling for a SOFC stack by using a Hammerstein model. Journal of Power Sources 175(1), 441–446 (2008)

76. Huoa, H.B., Zhua, X.J., Hub, W.Q., Tua, H.Y., Li, J., Yangd, J.: Nonlinear model predictive control of SOFC based on a Hammerstein model. Journal of Power Sources 185(1), 338–344 (2008)

77. Iwamoto, M., Williams, A., Chen, P.F., Metzger, A., Larson, L., Asbeck, P.: An extended Doherty amplifier with high efficiency over a wide power range. IEEE Transactions on Microwave Theory and Techniques 49(12), 2472–2479 (2001)

78. Jansen, M., Oonincx, P.J.: Second generation wavelets and applications. Springer-Verlag, London (2005)

79. Jeng, J.C., Huang, H.P.: Nonparametric identification for control of MIMO Hammerstein systems. Industrial and Engineering Chemistry Research 47(17), 6640–6647 (2008)

80. Jeraj, J., Mathews, V.: A stable adaptive Hammerstein filter employing partial orthogonalization of the input signals. IEEE Transactions on Signal Processing 54(4), 1412–1420 (2006)

81. Juditsky, A., Hjalmarsson, H., Benveniste, A., Delyon, B., Ljung, L., Sjoberg, J., Zhang, Q.H.: Nonlinear black-box models in system-identification - mathematical foundations. Automatica 31(12), 1725–1750 (1995)

82. Jurado, F.: A method for the identification of solid oxide fuel cells using a Hammerstein model. Journal of Power Sources 154(1), 145–152 (2006)

83. Jyothi, S.N., Chidambaram, M.: Identification of Hammerstein model for bioreactors with input multiplicities. Bioprocess Engineering 23(4), 323–326 (2000)

84. Kamiński, W., Strumiłło, P.: Kernel orthonormalization in radial basis function neural networks. IEEE Transactions on Neural Networks 8(5), 1177–1183 (1997)

85. Kelly, S., Kon, M., Raphael, L.A.: Pointwise convergence of wavelet expansions. Bulletin of The American Mathematical Society 30(1), 87–94 (1994)

86. Keys, R.: Cubic convolution interpolation for digital image processing. IEEE Transactions on Acoustics, Speech and Signal Processing 29(6), 1153–1160 (1981)

87. Kim, J., Konstantinou, K.: Digital predistortion of wideband signals based on power amplifier model with memory. Electronics Letters 37(23), 1417–1418 (2001)

88. Kim, W.J., Cho, K.J., Stapleton, S., Jong-Heon, Kim: Piecewise pre-equalized linearization of the wireless transmitter with a Doherty amplifier. IEEE Transactions on Microwave Theory and Techniques 54(9), 3469–3478 (2006)

89. Knuth, D.E.: The Art of Computer Programming. Volume 3. Sorting and searching. Addison-Wesley Longman Publishing Co., Inc, Boston, MA (1998)

90. Krim, H., Tucker, D., Mallat, S., Donoho, D.: On denoising and best signal representation. IEEE Transactions on Information Theory 45(7), 2225–2238 (1999)

91. Krzyżak, A., Linder, T.: Radial basis function networks and complexity regularization in function learning. IEEE Transactions on Neural Networks 9(2), 247–256 (1998)

92. Kukreja, S., Kearney, R., Galiana, H.: A least-squares parameter estimation algorithm for switched Hammerstein systems with applications to the VOR. IEEE Transactions on Biomedical Engineering 52(3), 431–444 (2005)

93. Lee, Y., Schetzen, M.: Measurement of the Wiener kernels of a non-linear system by cross-correlation. International Journal of Control **2**, 237–254 (1965)
94. Lloyd, S.: Least squares quantization in PCM. IEEE Transactions on Information Theory **28**(2), 129–137 (1982)
95. Lyons, R.: Understanding digital signal processing. Prentice Hall PTR (2004)
96. Mallat, S.G.: A Wavelet Tour of Signal Processing. Academic Press, San Diego (1998)
97. Marmarelis, V.Z.: Nonlinear dynamic modeling of physiological systems. IEEE Press Series on Biomedical Engineering. Wiley-IEEE Press, Piscataway, NJ (2004)
98. Mason, J.C., Handscomb, D.C.: Chebyshev polynomials. Chapman & Hall/CRC, Boca Raton (2003)
99. Max, J.: Quantizing for minimum distortion. IRE Transactions on Information Theory **6**(1), 7–12 (1960)
100. Meijering, E.: A chronology of interpolation: From ancient astronomy to modern signal and image processing. Proceedings of the IEEE **90**(3), 319–342 (2002)
101. Meilera, M., Schmida, O., Schudya, M., Hoferb, E.: Dynamic fuel cell stack model for real-time simulation based on system identification. Journal of Power Sources **176**(2), 523–528 (2008)
102. Morgan, D., Ma, Z., Kim, J., Zierdt, M., Pastalan, J.: A generalized memory polynomial model for digital predistortion of RF power amplifiers. IEEE Transactions on Signal Processing **54**(10), 3852–3860 (2006)
103. Mosteller, F., Tukey, J.W.: Data Analysis and Regression: A Second Course in Statistics. Addison-Wesley Series in Behavioral Science: Quantitative Methods. Addison-Wesley, Reading, MA (1977)
104. Mzyk, G.: Nonlinearity recovering in Hammerstein system from short measurement sequence. IEEE Signal Processing Letters **16**(9), 762–765 (2009)
105. Mzyk, G.: Parametric versus nonparametric approach to Wiener systems identification. In: F. Giri, E.W. Bai (eds.) Block-oriented nonlinear system identification, Lecture Notes in Control and Information Sciences, pp. 111–125. Springer (2010)
106. Nadaraya, E.A.: On estimating regression. Theor. Probability Appl. **9**, 141–142 (1964)
107. Nešić, D., Mareels, I.M.Y.: Dead-beat control of simple Hammerstein models. IEEE Transactions on Automatic Control **43**(8), 1184–1188 (1998)
108. Nordsjo, A., Zetterberg, L.: Identification of certain time-varying nonlinear Wiener and Hammerstein systems. IEEE Transactions on Signal Processing **49**(3), 577–592 (2001)
109. Ogden, R.: Essential Wavelets For Statistical Applications and Data Analysis. Birkhäuser, Boston (1997)
110. Pawlak, M., Hasiewicz, Z.: Nonlinear system identification by the Haar multiresolution analysis. IEEE Transactions on Circuits and Systems I: Fundamental Theory and Applications **45**(9), 945–961 (1998)
111. Pawlak, M., Hasiewicz, Z., Wachel, P.: On nonparametric identification of Wiener systems. IEEE Transactions on Signal Processing **55**(5), 482–492 (2007)
112. Pawlak, M., Rafajłowicz, E., Krzyżak, A.: Postfiltering versus prefiltering for signal recovery from noisy samples. IEEE Transactions on Information Theory **49**(12), 3195–3212 (2003)
113. Press, W.H., Flannery, B.P., Teukolsky, S.A., Vetterling, W.T.: Numerical Recipes in C: The Art of Scientific Computing. Cambridge University Press, Cambridge (1993)
114. Raab, F., Asbeck, P., Cripps, S., Kenington, P., Popovic, Z., Pothecary, N., Sevic, J., Sokal, N.: Power amplifiers and transmitters for RF and microwave. IEEE Transactions on Microwave Theory and Techniques **50**(3), 814–826 (2002)
115. Rabiner, L.: Multirate Digital Signal Processing. Prentice Hall PTR (1996)
116. Rafajłowicz, E.: Nonparametric orthogonal series estimators of regression: A class attaining the optimal convergence rate in l_2. Statistic & Probability Letters **5**, 219–224 (1987)
117. Rafajłowicz, E.: Consistency of orthogonal series density estimators based on grouped observation. IEEE Transactions on Information Theory **43**, 283–285 (1997)
118. Roger L. Claypoole, J., Davis, G.M., Sweldens, W., Baraniuk, R.G.: Nonlinear wavelet transforms for image coding via lifting. IEEE Transactions on Image Processing **12**(12), 1449–1459 (2003)

119. Rudin, W.: Principles of mathematical analysis, 3rd edn. McGraw-Hill, New York (1976)
120. Rutkowski, L.: Generalized regression neural networks in time-varying environment. IEEE Transactions on Neural Networks **15**(3), 576 – 596 (2004)
121. Rutkowski, L., Rafajłowicz, E.: On optimal global rate of convergence of some nonparametric identification procedures. IEEE Transactions on Automatic Control **34**(10), 1089–1091 (1989)
122. Said, A., Pearlman, W.A.: A new, fast, and efficient image codec based on set partitioning in hierarchical trees. IEEE Transactions on Circuits and Systems for Video technology **6**(3), 243–251 (1996)
123. Schetzen, M.: Nonlinear system modeling based on the Wiener theory. Proceedings of the IEEE **69**(12), 1557–1573 (1981)
124. Schilling, R.J., Jr., J.J.C., Al-Ajlouni, A.F.: Approximation of nonlinear systems with radial basis function neural networks. IEEE Transactions on Neural Networks **12**(1), 1–15 (2001)
125. Shapiro, J.M.: Embedded image coding using zerotrees of wavelet coefficients. IEEE Transactions on Signal Processing **41**(12), 3445–3462 (1993)
126. Shi, K., Zhou, G., Viberg, M.: Compensation for nonlinearity in a Hammerstein system using the coherence function with application to nonlinear acoustic echo cancellation. IEEE Transactions on Signal Processing **55**(12), 5853–5858 (2007)
127. Sjoberg, J., Zhang, Q.H., Ljung, L., Benveniste, A., Delyon, B., Glorenec, P.Y., Hjalmarsson, H., Juditsky, A.: Nonlinear black-box modeling in system-identification - a unified overview. Automatica **31**(12), 1691–1724 (1995)
128. Skubalska-Rafajłowicz, E.: Pattern recognition algorithms based on space-filling curves and orthogonal expansions. IEEE Transactions on Information Theory **47**(5), 1915–1927 (2001)
129. Śliwiński, P.: Fast algorithms for non-linearity recovering in Hammerstein systems with ordered observations. In: Proceedings 10th IEEE International Conference on Methods and Models in Automation and Robotics – MMAR 2004, pp. 451–456. Institute of Control Engineering, Technical University of Szczecin, Miedzyzdroje (2004)
130. Śliwiński, P.: On-line wavelet estimation of Hammerstein system nonlinearity. International Journal of Applied Mathematics and Computer Science **20**(3), 513–523 (2010)
131. Śliwiński, P., Hasiewicz, Z.: Computational algorithms for multiscale identification of nonlinearities in Hammerstein systems with random inputs. IEEE Transactions on Signal Processing **53**(1), 360–364 (2005)
132. Śliwiński, P., Hasiewicz, Z.: Computational algorithms for wavelet identification of nonlinearities in Hammerstein systems with random inputs. IEEE Transactions on Signal Processing **56**(2), 846–851 (2008)
133. Śliwiński, P., Rozenblit, J., Marcellin, M.W., Klempous, R.: Wavelet amendment of polynomial models in nonlinear system identification. IEEE Transactions on Automatic Control **54**(4), 820–825 (2009)
134. Slud, E.: Entropy and maximal spacings for random partitions. Zeitschrift für Wahrscheinlichtstheorie und vervandte Gebiete **41**(4), 341–352 (1978)
135. Srinivasan, R., Rengaswamy, R., Narasimhan, S., Miller, R.: Control loop performance assessment — Hammerstein model approach for stiction diagnosis. Industrial & Engineering Chemistry Research **44**(17), 6719–6728 (2005)
136. Steinhaus, H.: Sur la division des corp materiels en parties. Bull. Acad. Polon. Sci. **C1. III**(IV), 801–804 (1956)
137. Stiles, B., Sandberg, I., Ghosh, J.: Complete memory structures for approximating nonlinear discrete-time mappings. Neural Networks, IEEE Transactions on **8**(6), 1397–1409 (1997)
138. Stone, C.J.: Optimal global rates of convergence for nonparametric regression. Annals of Statistics **10**(4), 1040–1053 (1982)
139. Sung, S.W.: System identification method for Hammerstein processes. Industrial and Engineering Chemistry Research **41**(17), 4295–4302 (2002)
140. Sureshbabu, N., Farrell, J.A.: Wavelet based system identification for non-linear control. IEEE Transactions on Automatic Control **44**(2), 412–417 (1999)

141. Sweldens, W.: The lifting scheme: A custom-design construction of biorthogonal wavelets. Applied and Computational Harmonic Analysis **3**(2), 186–200 (1996)
142. Sweldens, W.: The lifting scheme: A construction of second generation wavelets. SIAM J. Math. Anal. **29**(2), 511–546 (1997)
143. Taubman, D., Marcellin, M.: JPEG2000. Image Compression Fundamentals, Standards and Practice, *The Kluwer International Series in Engineering and Computer Science*, vol. 642. Kluwer Academic Publishers (2002)
144. Thévenaz, P., Blu, T., Unser, M.: Interpolation revisited. IEEE Transactions on Medical Imaging **19**(7), 739–758 (2000)
145. Unser, M., Daubechies, I.: On the approximation power of convolution-based least squares versus interpolation. IEEE Transaction on Signal Processing **45**(7), 1697–1711 (1997)
146. Walter, G.G.: Pointwise convergence of wavelet expansions. Journal of Approximation Theory **80**(1), 108–118 (1995)
147. Walter, G.G., Shen, X.: Wavelets and other orthogonal systems with applications, 2nd Ed. Chapman & Hall, Boca Raton (2001)
148. Wang, J., Wang, D., Moore, P., Pu, J.: Modelling study, analysis and robust servo control of pneumatic cylinder actuator systems. IEE Proceedings: Control Theory and Applications **148**, 35–42 (2001)
149. Watson, G.S.: Smooth regression analysis. Sankhya, Series A **26**, 355–372 (1964)
150. Westwick, D., Kearney, R.: Nonparametric identification of nonlinear biomedical systems, part i: Theory. Critical reviews in biomedical engineering **26**(3), 153 (1998)
151. Westwick, D.T., Kearney, R.E.: Separable least squares identification of nonlinear Hammerstein models: Application to stretch reflex dynamics. Annals of Biomedical Engineering **29**(8), 707–718 (2001)
152. Westwick, D.T., Kearney, R.E.: Identification of nonlinear physiological systems. IEEE Press Series on Biomedical Engineering. Wiley-IEEE Press, Piscataway (2003)
153. Wiener, N.: Nonlinear problems in random theory. The MIT Press, Cambridge, Massachusetts, USA (1966)
154. Zhao, W.X., Chen, H.F.: Recursive identification for Hammerstein system with ARX subsystem. IEEE Transactions on Automatic Control **51**(12), 1966–1974 (2006)
155. Zhou, D., DeBrunner, V.E.: Novel adaptive nonlinear predistorters based on the direct learning algorithm. IEEE Transactions on Signal Processing **55**(1), 120–133 (2007)

References

Index